T0299233

Lowering Your Facility's Electric Rates

Lowering Your Facility's Electric Rates

by Lindsay Audin CEM CEP LEED AP

Routledge
Taylor & Francis Group

LONDON AND NEW YORK

Published 2020 by River Publishers

River Publishers

Alsbjergvej 10, 9260 Gistrup, Denmark

www.riverpublishers.com

Distributed exclusively by Routledge

4 Park Square, Milton Park, Abingdon, Oxon OX14 4RN

605 Third Avenue, New York, NY 10017, USA

Library of Congress Cataloging-in-Publication Data

Names: Audin, Lindsay, author.
    Title: Lowering your facility's electric rates / by Lindsay Audin, CEM CEP
LEED AP.
Description: Lilburn, GA : Fairmont Press, [2017] I Includes bibliographical
references and index.
Identifiers: LCCN 2017021471 (print) I LCCN 2017031276 (ebook) I ISBN
    9780881737820 (Electronic) I ISBN 088173781X (alk. paper) I ISBN
    9788770223386 (electronic) I ISBN 9781138303140 (Taylor & Francis
    distribution : alk. paper)
Subjects: LCSH: Cost control. I Electric power consumption--Costs. I Electric
    utilities--Rates. I Facility management. I Buildings--Energy conservation.
Classification: LCC HD47.3 (ebook) I LCC HD47.3 A93 2017 (print) I DDC
    658.15/53--dc23
LC record available at https://lccn.loc.gov/2017021471

Lowering your facility's electric rates /Lindsay Audin
First published by Fairmont Press in 2018.

Routledge is an imprint of the Taylor & Francis Group, an informa business

ISBN 088173781X   (The Fairmont Press, Inc.)
ISBN 9781138303140   (print)
ISBN 9788770223386   (online)
ISBN 9781003151302   (ebook master)

While every effort is made to provide dependable information, the publisher, authors, and
editors cannot be held responsible for any errors or omissions.

The views expressed herein do not necessarily reflect those of the publisher.

Acknowledgements and Dedication

While developing the webinars on which this book is based, I examined many publications covering the same turf. All were either far too basic, or far too theoretical. As a result, I originated most of the material, based on experiences of my own and those of my associates.

Thank you to all the following for allowing me to re-purpose various materials I wrote for them in the past:

- New York State Energy Research and Development Authority (NYSERDA)

- Noesis.com

- Association of Energy Engineers (AEE)

Thank you also to the following for allowing me to use materials from their publications:

- Platts.com / Megawatt Daily

- Power Marketers Association

Thank you to Sondra Audin Armer for all her indexing assistance, and thank you to all the federal and state agencies and commissions for access to their vast troves of public materials, orders, rulings, proceedings, etc. referenced in this book.

Some of the techniques covered herein are expansions of one-off ventures that later became templates used by practitioners for their proprietary energy services. I acknowledge their contributions to the field, and to them I dedicate this book. Thanks to all for your cleverness and willingness to share some of your best tricks.

If readers have any others they care to share, send them to me (energywiz@energywiz.com) and I'll put them in the next edition. We can't keep all these goodies to ourselves.

Table of Contents

SECTION I: BACKGROUND

Chapter 1
The Bottom Line 3

Chapter 2
Understanding The Basic Power System 5
    Evolution of the Electric Utilities 5
    Utility Geography 7
    Utility Financial Structures Vary. 8
    New Players in the Power Game 9
    How the Grids Expanded 10
    The Three Main Functions Of Utilities 12
    Retail Deregulation Impacts Some Utilities 13

Chapter 3
Understanding Power Pricing 17
    Average Versus Actual Electric Rates 17
    To Understand the Options, Learn the Lingo 19
    Unbundling Electric Rates 20
    Basic Electricity and Tariff Units. 20
    How Peak Demand May Be Measured 24
    Why Understanding Peak Demand Is Important 26

Chapter 4
Typical Components in a C&I Electric Bill. 29
    Fixed Charges 29
    Major Variable Charges. 30
    Program Charges and Taxes 32
    Many Pieces in the Pie 33

Chapter 5
First Steps in a Rate Analysis 35
    Tips to Make Analysis Easier .36
    Break Down Dense Language 37
    Use Utility Accounting Software 38

Review Alternative Rates. .39
What Tariffs May Not Show 40

SECTION II: OPTIONS FOR RATE AND COST REDUCTIONS

Chapter 6
(Mostly) Administrative Options 45

Avoid, Shift, and / or Reduce Taxes 45
Minimize Bill Processing Costs 47
Aggregate Meters / Accounts. 48
Re-classify Loads / Accounts 50
Lower Service Firmness 52
Seek A CHP Deferral Option. 53
Find Billing Errors Using a Professional Utility
    Bill Auditor 54
Utilize Financial Techniques 56

Chapter 7
Utilizing Metering and Dynamic Rate Options. 59

How Utility Metering Is Changing 61
A Brief Tutorial on Interval Data 62
How to Get the Data 66
Evaluating a TOU Rate. 68
Power Pricing Based on Utility Load Profiles 70
A Slightly Extreme Example .71
Who May Benefit from TOU? 72
Get It 'Wholesale' via RTP 73
Who May Benefit from RTP?. 76

Chapter 8
Equipment/Technical/ Operational Options 77

Reduce or Eliminate a Reactive Power Charge 77
Convert Electric Loads to Natural Gas 78
Get Paid to Cut Load By Demand Response Programs. 80
What DR is All About 80
Who Offers DR Programs / Tariffs? 83
Cut a Distribution Rate by Upgrading Voltage Service. 83
Pursue Transmission Bypass. 86
Buy Cheaper Power Generated On-site. 87

Negotiating Cogeneration 88
Cash in on Options for Using Alternative Energy Sources. . . .89

SECTION III: CHANGING THE RULES

Chapter 9
Negotiating with Utilities. **95**
IOUs: Often the Toughest Nuts to Crack. 96
Munis and Co-ops May be Easier. 97
Authorities Are a Mix. 98
Possible Mechanisms for a Resolution 98
Negotiation Is Never a Simple or Easy Task. 99
Sweat the Details of a Proposed Deal. 101

Chapter 10
Intervening to Help Write the Rates **103**
Introduction 103
Understanding Rate Proceedings 104
To Succeed, Don't Go It Alone. 105

SECTION IV: COMPETITIVE RETAIL
POWER PROCUREMENT

Chapter 11
Buying from Competing Electricity Suppliers. **111**
Introduction 111
Historical Background 112
How the Purchasing Process Works. 114
Preparing for Retail Power Procurement. 116

Chapter 12
Options for Competitive Supply Pricing. **123**
Price Content and Structure 123
Common Types of Power Products 126

Chapter 13
Managing Competitive Power Procurement. **131**
Finding and Choosing Retail Power Suppliers 131
RFI Content 131
Price Discovery Methods and Tools. 134
RFP Content 138

Chapter 14
Ways to Save Money and/or Increase Value 141

Chapter 15
Negotiating/Improving a Supply Contract 157
    How to Improve a Contract 159
    Post Bid Options 162

Appendices 165

Glossary of Tariff and Energy Procurement Terms 219

Index 391

Introduction

While teaching classes in electric load profiling and power procurement, I found that many energy practitioners were uninformed (or misinformed) regarding how electricity is actually priced. Many were also unaware of how to properly calculate potential dollar savings from energy efficiency upgrades.

Halfway into a discussion of minimizing ratchet charges, for example, a participant would ask: "Now, tell me again: what's the difference between a kilowatt and a kilowatt-hour?" At that point, I realized that most of what I had already covered had gone over that person's head.

But every problem is merely an opportunity flipped on its back. In this case, there was obviously a need to backfill that knowledge vacuum with basic information. To do so, I created several on-line courses on electric rates, load profiling, and purchasing of competitively-priced power.

Designed for commercial/industrial/institutional (C&I) power customers, this book details ways to secure lower electric rates and pricing in both regulated and deregulated retail power markets. The range of options varies from simple methods (e.g., minimizing sales tax) to more complex techniques (e.g., intervening in regulatory rate proceedings).

As with any technical undertaking, electricity pricing has its own lingo and abbreviations. The extensive glossary in appendix J translates acronyms and price-related terminology (both retail and wholesale). The other appendices flesh out details covered in the body text.

During my 40+ years of experience in energy services, I pursued (or worked with others) to pursue the options covered in this book. My first-hand experiences as energy manager, consultant, buyer, broker, tariff negotiator, and customer rep revealed to me the many pros and cons of those options.

Here's your first tip: Don't be overwhelmed the first time you try to read a tariff or power contract. With a little patience and parsing, each may be translated into common English and spreadsheet formulae. Once you clearly understand your power pricing and ways to reduce it, others who don't will think you're a genius.

Acronyms and Abbreviations

B&E = blend-and-extend; a method to capture future potential savings when a forward curve is falling but an existing contract has not yet run out

CHP = combined heat and power, also known as cogeneration, wherein an on-site power plant provides both forms of energy

C&I = commercial and industrial

DAM/HAM = day-ahead/hour-ahead markets (real-time pricing)

DR = demand response (program to pay those that cut load upon request)

DSM = demand-side management

EDI = electronic data interchange

EFT = electronic fund transfer (efficient way to pay electric bills)

ESCo = Energy Service Company (finances, installs, operates efficiency upgrades; In New York state, this also means a power supplier)

EWG = exempt wholesale generator (power producer that can't sell to end use customers)

FERC = Federal Energy Regulatory Comm. (oversees interstate transmission)

FIT = Feed-In Tariff (potentially lucrative option to sell on-site PV power to utility)

HVAC = heating, ventilation, and air conditioning system

ICAP = installed capacity (LICAP = locational ICAP)

IDD = Industrial Development District (where special tax breaks or other incentives, e.g., subsidized electric rates may exist)

IM = interval meter

IOU = investor-owned utility (regulated monopoly owned by shareholders)

ISO = independent system operator (public body managing regional transmission grid)

IPP = independent power provider (a form of non-utility generator)

LDC = local distribution company (e.g., your local utility)

LF = load factor (average demand/peak demand)

LMP = locational marginal pricing (supply rate based on a zone within an ISO)

LSE = load-serving entity (a power supplier or utility)

M&V = measurement and verification

NUG = non-utility generator

OTC = over the counter (i.e., wholesale energy trades outside a regulated exchange)

PMA = Power Marketing Administration (federal agency selling wholesale power)

PP = power pool (agency in a region where utilities coordinate transmission and generation)

PPA = power purchasing agreement under which a customer buys power from an on-site developer or third-party supplier

PUC = public utility commission (a government agency that creates and enforces regulations on behalf of consumer interests)

PUD = Public Utility District (name/region defining some types of co-op utilities)

PV = photovoltaic solar panels that may be located at a customer's site

QF = qualifying facility (specialized category of cogenerating or renewable power source)

ROE = return-on-equity (a rate at which a utility may recover its costs based on its investments)

ROI = return-on-investment (approximately the inverse of payback period)

RTO = regional transmission operator (an ISO with expanded powers)

RTP = real-time pricing (e.g., based on hourly wholesale grid pricing)

T&D = transmission & distribution (together, these may be called "delivery")

TOU = time-of-use (energy pricing differentiated by time)

UDC = utility distribution company (same as LDC)

Section I: Background

Chapter 1

The Bottom Line

There are essentially 4 ways to cut a facility's average electric *rate* (as versus its electric *consumption*):

1) *reduce regulated tariff rates* for supply and/or delivery

2) *minimize peak kW demand*, both monthly and annually

3) *capture incentives and credits* offered in utility and/or government programs, including tax breaks

4) *secure cheaper supply* via subsidies, competitive procurement, and/or self-generation.

Customers in states that allow competitive retail power procurement may choose any (or all) of the above. Others may find most (but not all) of those options available to them.

No one technique is likely to cut your electric bills by more than a few percent. Pursuing multiple options, however, may together do so by 10% to 20% at a lower cost than most energy efficiency measures. While nothing should discourage pursuit of the latter, they should be developed in concert with rate-cutting options that may involve little or no financial investment. In some cases, merely filing appropriate paperwork may yield savings. In others, a more sophisticated and/or aggressive approach may be necessary.

For some customers, no savings may be possible. For others, potentially lucrative rewards have made the effort worthwhile for those having the patience and persistence needed to achieve them.

Key to such success is parsing and understanding each cost component of your electricity, how it is regulated (or competitive), and ways to lower its rate.

Here's another tip: the best time to minimize electric rates is before opening a new electric account for a facility. Steps taken during the design phase for a new building, wing, or plant upgrade may yield significant savings that may be difficult or impossible once construction starts. While it's possible to fix errors later, doing so may cost more than taking a pro-active approach. Several of the options discussed herein (e.g., securing high-tension electric service, raising power factor) are easier to accomplish as part of design and specification than later via negotiations or lawsuits.

Some options require a grasp of issues that may be beyond the job descriptions of many facility (and some energy) managers. To assist those needing a broader background in such matters, please see the appendices.

Chapter 2

Understanding
The Basic Power System

A variety of organizations and entities are involved in producing and delivering power in the U.S. The description that follows is a general overview. For more detailed information, consult the web sites found in the text. To focus the discussion and avoid issues not relevant to the purpose of this book, I have distilled some details and history to concentrate on those that relate to retail power pricing.

EVOLUTION OF THE ELECTRIC UTILITIES

America's electric power system grew in an uncoordinated fashion over several decades. Urban areas were the first to be electrified, with most local systems privately owned. In rural areas, and in cities either too small or where electric service was not profitable, publicly owned systems developed with the help of federal, state, and municipal governments. Regardless of ownership, most systems were isolated entities that built and operated generation to satisfy their local loads.

In some cases, that isolation created incompatible power systems. Until 1936, Southern California Edison (a privately owned utility) was still running on 50-cycle (Hz) current, while most of the nation used 60-Hz power. The adjacent Los Angeles city-owned utility did not convert to 60 Hz until 1948. For many years, those using 50-Hz power had to buy special clocks and other devices because standard models ran at a different speed.

Before utilities were regulated, private power companies competing for customers installed power systems with overlapping wiring, yielding high costs and a chaotic situation. To deal with the problem and to encourage economic development, states stepped in with a regulatory compact: they would grant utilities exclusive monopolies in geographically defined territories in exchange for the right to regulate their pricing, services, and other activities. By 1916, 33 state regulatory agencies had been established. By World War II, all other states had created their own. Generically called Public Utility Commissions (PUCs), they oversee and regulate many aspects of retail power services.

Four different types of utilities operate in the U.S.:

- Investor-Owned Utility (IOU): Owned by stockholders, they often have complex and varied rate structures. About 72% of U.S. customers get their power through IOUs.

- Co-operative Utility (Co-op): Owned by ratepayers, they are common in rural areas. Tariffs are sometimes simpler than those for IOUs. They supply power to about 11% of U.S. customers.

- Municipal Utility (Muni): Owned by city or county agencies or governments, some have the simplest tariffs. They serve about 16% of U.S. customers.

- Authorities or Power Marketing Administrations (PMA): Owned by federal or state-run agencies, they make and sell wholesale power to co-ops, munis, economic development agencies, some IOUs, and occasionally to large retail customers. Their tariffs may mirror those of IOUs, and may be limited to wholesale supply at high voltage rather than delivery to end users.

Because IOUs have stockholders, such utilities are privately owned. The other three types are owned or chartered by governments, and are publicly owned.

IOUs are regulated by PUCs, while publicly owned utilities

may regulate their own pricing through boards elected by rate-payers or chosen by local government bodies, but with some state oversight.

Over 4,000 utilities exist in the U.S. (see Figure 2-1). The light bars show how many of each type exist, while the dark bars show how many GW each generates as a group (1 GW = 1,000 MW). Note that most power supplied by co-ops and munis is not generated by them. Instead, they may purchase it from IOUs, authorities, or various types of privately owned "merchant generators" (see "other" in the chart) described below. Owning about as much generating capacity as all IOUs, munis, and co-ops combined, they are now major players in the U.S. electricity industry.

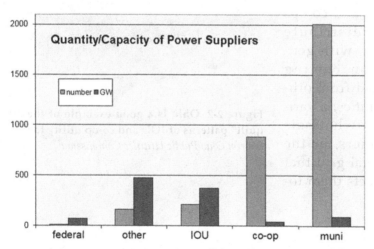

Figure 2-1. Light grey bars indicate the quantity of each type of utility/supplier. Dark bars indicate the gigawatt (GW) capacity of their power plants (1 GW = 1,000 MW). "Other" is private wholesale merchant generators. *Source: author.*

UTILITY GEOGRAPHY

The growth of utilities over decades created a patchwork of territories, often with complex boundaries. Ohio is a good example (see Figure 2-2). It contains:

- 7 IOUs (the large contiguous regions)

- 25 co-ops (patchy odd-shaped spaces)

- 82 munis (with territories too small to show here)

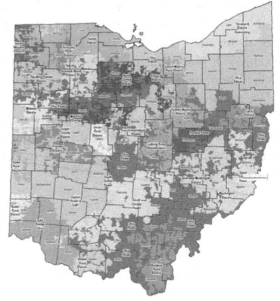

Each of those 114 utilities has its own power pricing schedule (called a "tariff."

Many of the munis and co-ops only distribute power, with generation coming instead from other utilities, a variety of wholesale suppliers, or the regional grid that connects them together.

Figure 2-2. Ohio is a good example of the "crazy quilt" patterns of IOU and co-op utility territories. *Source: Ohio Public Utilities Commission*

UTILITY FINANCIAL STRUCTURES VARY

The regulatory compact with IOUs limits their return-on-equity (ROE) for money invested in construction and acquisition of power plants, transmission, etc. (what some utility people call their "iron in the ground"), though efforts are afoot to stretch that to include software and other investments. That ROE ranges from about 9% to 12%, as allowed by each PUC. If, for example, an IOU spent $1,000 on a transformer, it could charge ratepayers

$90 to $120 a year for it. The cost of the transformer would then be recouped in 8 to 10 years. In each year thereafter, the utility would earn another $90 to $120 as long as the transformer was in operation or storage. Since it might last 40 years, a significant financial return may accrue during the remainder of its lifetime. This process encouraged growth in utility systems and the electrification of much of the nation.

Financing of publicly owned utilities was instead aided by government funding and borrowing. Many co-ops, for example, benefited from rural flood control efforts that created dams which also generate vast amounts of power, most of which is sold to publicly owned utilities. Through government bonds and other low-interest financial products, such utilities may secure capital at lower rates than IOUs. Without needing to pay dividends to shareholders, many (but not all) have been able to provide power at lower rates than nearby IOUs.

NEW PLAYERS IN THE POWER GAME

For most of the 20th century, utilities were the only entities making and selling power to customers lacking their own generation. In 1978, the Public Utility Regulatory Policies Act (PURPA) opened the door to non-utility generators (NUG) by creating the qualifying facility (QF) category. It was initially focused on cogeneration, now called "combined heat and power" (CHP) systems. Units up to 80 MW used fossil fuel more efficiently than utility power plants. The owners of such facilities were designated as "independent power producers" (IPP), though that term is now also used to describe other types of non-utility power suppliers.

A QF may also produce power from renewable sources (e.g., wind, solar). As renewable power technologies advanced, they eventually became a common form of QF. With a few exemptions, QF power was sold only to utilities. Utilities were required to buy QF-generated power under long-term contracts based on their avoided cost for the next kWh they could either generate or purchase.

The Energy Policy Act of 1992 (EPAct) changed some of those rules. It created yet another NUG category, that being "exempt wholesale generators" (EWG) that can only sell power to a wholesale grid or utility, without restriction as to how it is generated. Designed as a first step toward deregulation of wholesale power, EWGs were guaranteed equal access to the transmission grid.

As power grids changed, QFs were pressed to become competitive. Wherever wholesale markets are considered "organized" (described below), utilities are no longer required to buy QF power at their avoided costs under long-term contracts.

HOW THE GRIDS EXPANDED

As utilities' territories expanded and began to border each other, some determined that it was economical to connect to each other as a low-cost way to enhance reliability and avoid the need to build and maintain completely independent systems. Over time, localized power systems interconnected via high-voltage transmission grids. Utilities were then able to support each other's systems and trade power as a way to minimize their operating costs. For decades, on an as-needed basis they transferred power between themselves with a minimal markup (e.g., 15%) over their own generating costs.

After the 1965 blackout that covered much of northeast U.S., grid connection standards were tightened. The Federal Energy Regulatory Commission (FERC)—at the time called the Federal Power Commission—began to play a more important role in maintaining system reliability.

To coordinate and maintain standards for all of them, the North American Electric Reliability Council (NERC), a private entity, was formed in 1968. It has 9 regional councils that, upon its formation, mirrored the limits of the regional grids existing at the time. A current map of their boundaries (and additional information) may be found at: www.nerc.com/AboutNERC/keyplay-ers/Pages/Regional-Entities.aspx Today's grids cover multiple states, connecting hundreds of utilities, and thousands of power plants.

While mostly stable, their boundaries sometimes change. At the time of this writing, for example, the Mountain West Transmission Group (which includes all of Colorado, most of Wyoming, and bits of several adjacent states) is considering joining (by 2019) the Southern Power Pool (SPP), adding more than 15,000 miles of transmission to it. Doing so would bring a variety of IOUs and publicly owned utilities into SPP's growing wholesale power market.

For most of the 20th century, utilities controlled transmission access and pricing to meet their own needs via private power pools (PP) representing their interests. Starting in 1992, federal rulings and laws began to change that landscape. FERC began chartering non-profit independent system operators (ISOs) that took over management of some regional grids. It also issued sweeping orders that allowed NUGs to use interstate transmission systems to move their power.

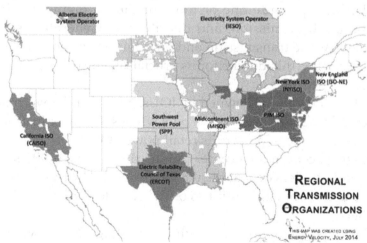

Figure 2-3. Regional "organized markets" controlled by FERC-chartered transmission organizations cover much of the US and parts of Canada. *Source: Federal Energy Regulatory Commission*

Unlike the early power pools, ISOs are run by boards and committees with representatives that—in addition to utilities—include NUGs, retail power suppliers (where states allow customers to buy from them), environmental and customer groups, various state agencies, and state public utility commissions (PUC). A current map of the ISOs may be found at: www.isorto.org/About/Role.

Grids controlled by ISOs are considered "organized markets." They manage wholesale power transactions that ensure sufficient hourly power is produced at the lowest cost to meet hourly loads via continuous competition among generators. As of 2017, almost 70% of both power customers and the miles of high-voltage transmission serving them are managed by some form of ISO (some are called regional transmission operators, or RTO). Most of the remaining non-ISO transmission exists in rural western and southern U.S.

THE THREE MAIN FUNCTIONS OF UTILITIES

Utilities provide one or more of the following services: generation, transmission, and distribution. Generators *supply* electricity, while transmission and distribution systems *deliver* it. Understanding this distinction is crucial when seeking ways to reduce electric rates.

Most generation is owned by IOUs, NUGs, and government agencies. Some produce and sell only wholesale power to utilities that later sell it to end users as retail power. Only very large customers (e.g., central data centers, industrial plants), utilities, and commodity traders deal at the wholesale level. The rest of us buy retail power from either utilities or (where allowed) retail power suppliers.

Power is *transmitted* through high-voltage lines owned by utilities and several transmission-only firms. Think of voltage as the "pressure" of power: a high-voltage transmission line is like a fire hose while a residential power line is like a garden hose.

Power plants may produce electricity at about 20,000 volts and step it up (using transformers) to the 100,000s of volts required to move it through transmission lines. Substation transformers at switching stations located along the transmission lines step down transmission voltage to 10,000s of volts.

Local *distribution* systems step that power down to 100s of volts through distribution transformers near (or at) a customer's facility. Utilities own the distribution system, and (with a few exceptions) its metering.

Utilities vary in the functions they provide. Instead of generating it, many utilities merely deliver power they get from others. Some do not even transmit power, instead taking it from substations owned by others and stepping it down for distribution on their local systems. Utilities that generate little or no power are called local distribution companies (LDC) or utility distribution companies (UDC).

Unless buying their supply from others, retail power customers pay their utility for all those services. The utilities then pay others for the services the utilities don't directly provide. Unaware of the interconnection of power systems and their trading processes, many customers falsely believe that their local utilities generate all the power they receive. As a result, some fear that closure of a nearby power plant could impair their access to electricity. That limited view may blind them to ways to reduce their electricity pricing.

RETAIL DEREGULATION IMPACTS SOME UTILITIES

Between 1996 and 2001, almost half the states considered ways to deregulate their IOUs so retail customers could buy electricity from non-utility suppliers. Some states eventually dropped the idea, and others suspended the effort after negative experiences, but 17 states and the District of Columbia pushed ahead. Via state legislation or PUC regulation, they required IOUs within their borders to separate their *supply* service from their *delivery* service (see the map in Figure 2-4).

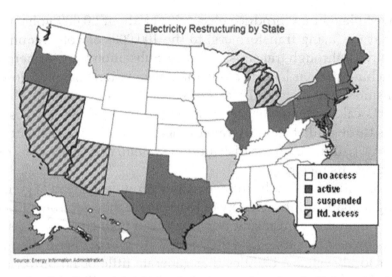

Figure 2-4. Retail electricity deregulation covers less than half the states, but much of their population. *Sources: US DOE and author.*

In most of those states, roughly 80% of commercial and industrial customers now buy their supply from competing electricity vendors, instead of their local utility. In those states, participation by residential customers has remained in the 20-30% range. In each case, a customer buys supply through a bilateral contract between it and a supplier, with no utility involvement. The power continues to be delivered by the local IOU under its *delivery* tariff.

No munis or co-ops were required to participate in retail deregulation. A handful have, however, allowed a few of their larger customers to gain access to the wholesale market through them, generally as an incentive to remain in their franchise area.

Despite various efforts, no other states have deregulated their IOUs since 2001. In a few states (e.g., Nevada), some loosening of regulatory constraints has occurred and may continue over time. In others (e.g., California), deregulation was suspended and/or severely limited. Barring major political changes, it is unlikely that other states will fully deregulate their retail electric supply markets. Options for securing deregulated supply are detailed in later chapters.

The rapid development of renewable power and desire by some customers to use it have, however, led to a few limited opportunities. Where utilities remain regulated, some IOUs have voluntarily chosen to allow a handful of large customers (primarily data centers) to buy wholesale renewable power from NUGs with whom those customers have purchasing contracts. The IOUs deliver (i.e., "wheel") that power only to them under so-called "green" tariffs. Elsewhere, some still-regulated states either allow or are considering ways that third-party owners of solar photovoltaic (PV) power plants may sell power to customers hosting that equipment on their property.

The first step in understanding what options may be available is determining the type of utility your facility is "behind," i.e., the utility territory in which it resides. Your utility's web site will likely clarify that issue. If it is behind a muni or co-op, or an IOU in a still-regulated state, sections of this book related to retail power procurement may not be relevant to you. Other sections, however, pertain to all commercial and industrial (C&I) customers regardless of utility type.

Figure 2-5. Retail deregulation affects only power generation. *Photo credit: US DOE.*

Chapter 3

Understanding Power Pricing

AVERAGE VERSUS ACTUAL ELECTRIC RATES

An average electric rate (i.e., total annual electric spend divided by total annual kWh) is a handy number for energy budgeting. It may, however, be inappropriate when calculating projected savings from an energy-related project, such as a lighting efficiency upgrade. An average rate may include rate components (e.g., fixed costs, demand charges, capacity charge) that may be unaffected by changing some types of equipment and/or occupant behavior.

Using an average rate as basis for a savings claim, payback analysis, or projected rate-of-return may then result in disappointment—or worse. Once credibility is lost on one project, it may be difficult to secure it for another. Understanding which rate components may be impacted by changes to a facility will guide you when calculating their financial impact—and (hopefully) keep you out of trouble.

When we buy electricity, we are purchasing a batch of energy commodities and services. The prices of each of those "rate components," are listed in documents called "tariffs" available from utilities and the regulatory bodies that oversee them. Each component may be parsed for examination and possible adjustment. Think of a tariff as your contract and price list with the utility. It details your rights, options, and limitations. Dissecting it may reveal options for reducing costs to your facility.

Utility tariffs for *retail* power are regulated by public utility commissions (PUC) or boards whose members may be appointed by governors or elected by voters. Tariffs for *wholesale* supply to

17

utilities are governed by the Federal Energy Regulatory Commission (FERC). Only very large (e.g., over a ~100 MW load) power customers such as small utilities must deal with FERC tariffs. We will instead focus on retail tariffs impacting customers that actually use the energy.

Tariffs are altered through proceedings overseen by PUCs. Parties to those actions typically include the utility, commission staff, large customers (and groups of them), lobbyists, government agencies, energy service companies, environmental groups, and a variety of trade and non-profit organizations. Many large customers use energy consultants to help them navigate this regulatory maze, and to find options for cutting pricing via alternative rates, altering their usage patterns, and special discounts.

Rate structures vary widely, but have similarities. While there are no hard and fast rules, my experience has found that IOU tariffs often have the most complex and varied structures. IOUs may also be resistant to negotiating options unless they are subsidized. PUCs grant them wide latitude on how they structure their rates, as long as they are developed through an open process that avoids favoring one group over others (unless a PUC has a reason to allow it).

Tariffs of publicly owned utilities may be simpler than those of IOUs. Co-ops may also be more open to working with customers through tariffs that encourage local economic development. Munis may have some of the simplest tariffs, and may be open to rates that help local businesses remain and/or grow in their territories. While relatively few retail customers will need to deal with federal or state-run power authorities, those that do (e.g., city governments) may find their tariffs parallel the simplicity of municipal and co-op utilities. Others may look more like those of nearby IOUs. Such authorities may also offer retail power to customers that maintain or expand employment, or otherwise bring economic development to their communities.

Electric rates are a continually moving target. While most changes are gradual, a few may seem sudden. California, for example, is (by 2019) moving all IOU customers to default rates

Figure 3-1. Standard mechanical meters (on the left) are being replace by digital meters (on the right) which measure and record electric usage in short (e.g., 15-minute) intervals and may communicate wirelessly with utilities and customers. *Photo credits: US DOE*

that vary with the time-of-use (TOU), unless a customer opts out. The installation of digital (so-called "smart") meters is opening the door to changes that may not have been feasible (or cost-effective) with mechanical meters, such as mandatory hourly pricing.

TO UNDERSTAND THE OPTIONS, LEARN THE LINGO

As previously discussed, the two most basic services provided by utilities are to supply and deliver electricity. As further explained in chapters addressing retail power procurement, all utilities deliver power but not all *supply* it.

"Supply" means to produce (i.e., generate and/or procure from others) electric energy (measured in kilowatt-hours [kWh]) and to provide the generating capacity (measured in kilowatts [kW]) needed to do so. When we take supply, we receive electric *energy* as kWh. In doing so, we are placing a load on a generator's kilowatt *capacity* (kW). That load is called "demand." *It is the speed at which we are consuming electric energy.* Both the kWh and the kW are measured by one or more electric meters for each of a facility's utility accounts.

Where retail electricity has been deregulated, the supply portion of tariffs may be replaced by private bilateral contracts between non-utility suppliers and their customers. Note, however, that retail deregulation is an ongoing process. Rules are still being written, and re-written, such that some supply pricing options may appear, change, or disappear over time.

Delivery of energy and capacity involves building and maintaining power lines, boosting and reducing voltage with transformers, ensuring consistent voltage and alternating current wave forms, metering and billing, etc. Some of those activities may be charged separately in a tariff. Delivery may constitute about 40% to 70% of a customer's total cost for electricity.

UNBUNDLING ELECTRIC RATES

While many utilities break down charges for those services in their tariffs, some (e.g., Nebraska's state-owned utilities) still do not (as of 2017). In such cases, it may be difficult to find ways to minimize them. Where such transparency is limited (or non-existent), I strongly suggest ratepayers push their regulatory bodies to differentiate tariff charges (called "unbundling"). Most rate-cutting tricks and techniques require an ability to address individual rate components. In this book, I have assumed that your rates have been at least partially unbundled.

Some utilities offer to provide upon request a "calculated" bill showing most (if not all) charges broken out for review. For an example and explanation, see Appendix H.

BASIC ELECTRICITY AND TARIFF UNITS

Becoming fluent with them is essential to understanding how to pursue this process.

- energy—capability to cause physical change (e.g., make something hotter or colder)

- power—how fast energy is made, moved, or consumed

- voltage (V)—the "pressure" of electricity

- amperage (A)—the flow rate of electricity

- kilowatt-hour (kWh)—a unit of electric energy, just as a therm is a unit of natural gas energy

- kilowatt (kW)—a unit of power, like horsepower (some utilities instead use kilovolt-amps [kVA])

- power factor (PF)—phase shift between volt and amp wave forms as used at your site (ideal PF is 1.0, or 100%); for single-phase electricity, PF = kW/kVA

To develop rates and charges, tariffs may use those technical terms in specific ways:

- consumption—quantity of kilowatt-hours (kWh) used in a billing period (in the U.S., that's typically a month)

- peak demand—highest speed of kWh use (expressed as kW or kVA) in a billing period; depending on the type of electric meter, it may be determined by the largest number of kWh used during a defined time interval (e.g., 15 minutes) within a given period (e.g., 8 AM to 6 PM)

- rate class—a grouping of customers typically based on their peak demand or annual kWh consumption, and/or by type, e.g., commercial, residential, industrial

- generation—ability to create power

- transmission—ability to move power at high voltage

- distribution—ability to move power and step down its voltage to levels used by customers (e.g., primary distribution may be for power delivered at voltages above 480 V, while secondary distribution occurs at or below 480 V)

- capacity—ability to make, move, or alter power

- on-peak—time when prices are highest, for both kWh and kW, e.g., 8 am-10 pm weekdays

- off-peak—time when prices are lowest; e.g., 10 pm-8 am, and maybe all day on weekends and holidays

- power factor—charge for quality of power use (also called reactive power) below a defined level, e.g., 95%; it may be charged as $/KVAR (i.e., kilovolt-amp reactive)

- fuel (or energy) adjustment charge (FAC)—cost of utility fuel and wholesale power above a base level defined in a tariff. A FAC makes retail power pricing sensitive to energy costs seen at wholesale.

- mill—one tenth of a cent (i.e., one thousandth of a dollar); power pricing may be seen in mills, e.g., 50 mills/kWh = 5 cents/kWh = $.05/kWh. One mill/kWh = $1.00/MWh.

- ratchet—monthly charge based on a percent of the highest monthly peak demand seen in the last year (which may be the prior calendar year, the prior 12 months, or a defined 12-month period)

Your utility has likely designated a representative for your account(s). Developing a cordial working relationship with him/her is a good first step when seeking help with tariff issues. If you don't understand a tariff term, ask that person to explain it to you.

An Easy Way To Lose Your Credibility

Many facility owners and managers (including the author) have successfully reduced their facilities' (or clients') electric bills through investments in energy-efficient equipment. Nothing in this book should be perceived as discouraging that valuable pursuit. But many do so without fully understanding how to calculate the value of the energy they expect to save.

A common mistake stems from using average electric pricing (i.e., annual spend divided by annual kWh). That number is NOT your electric rate: it is merely your *average* price. The following example should clarify just how wrong that could be.

A facility manager of an office building is considering installation of occupancy sensors that turn off lights in a room when it's unoccupied. To secure funding for such an upgrade, the facility requires that it earn at least a 20% rate-of-return (i.e., up to about a 5-year simple payback period). At the facility's *average* electric price of $.12/kWh, the payback period works out to about 4 years, so the project gets the green light.

A review of the electric tariff, however, reveals that average price includes these costs:

- a peak demand charge of $7/kW-month

- a monthly ratchet charge of $4 per peak annual kW demand

- a base charge of $400 a month, regardless of how much power is consumed

- a tiered time-of-use (TOU) block energy charge wherein the incremental cost of an avoided kWh is $.07/kWh during on-peak periods and $.05/kWh during off-peak periods.

Occupancy sensors are great at shutting off lights at night and on weekends, but may not significantly reduce the peak demand or consumption that occurs when most spaces are occupied. They also won't cut a base monthly charge, nor an annual ratchet charge. If they operate primarily during the tariff's off-peak period (e.g., 8 PM to 8 AM), most saved energy in this example may be worth only $.05/kWh.

When properly calculated, that projected 4-year payback period then balloons to over 9 years, of which the Chief Financial Officer (CFO) would never have approved. Fortunately for many energy and facility managers, few projects (unless quite large) are ever subjected to the measurement and verification (M&V) scrutiny that would reveal such blunders.

In some cases, however, a consultant may be called in to examine why the money spent on efficiency hasn't cut electric bills as much as promised. I've done so many times. I had the unfortunate duty of informing a CFO that his facility manager may have been conned by a vendor's self-serving savings calculations, or otherwise failed to perform appropriate due diligence. As a result, the manager lost credibility and (in a few cases) his position.

HOW PEAK DEMAND MAY BE MEASURED

Crucial to finding dollar savings and/or refunds is a grasp of how peak demand is determined and billed. Unlike electric rates for homes, rates for commercial and industrial customers may be based not only on how *much* electricity is consumed (i.e., kWh) but also on how *fast* it is consumed (i.e., kW). In the near future, it may also be partially based on *when* it is consumed.

Never confuse kW with kWh! Many (especially those in commercial real estate) still say "kilowatts" when they mean "kilowatt-hours." In my experience, doing so is a sure sign of tariff ignorance. See the boxed story below for a tale of someone who did, and almost lost $120,000 in the process.

Peak demand may be measured in several ways. Until the mid-2020s, by which time most mechanical meters may have been replaced, customers are likely to find one of three types of utility billing meters, each of which determines demand differently.

A *thermal* meter contains a metal coil which expands when heated by electric current being measured by the meter. A relatively consistent load for 15-20 minutes causes the unwinding coil to move a mechanical pointer. When the coil cools, the pointer remains in its highest position, setting a peak kW, regardless of the time at which it occurred. The pointer must be manually reset by the meter reader each month. If that does not occur, the customer may find subsequent demand charges are based on a recurring number unrelated to actual operation. Such excess demand charges are a common billing error. No data is stored in such meters.

A *pulse* meter contains a spinning rotor whose speed may be proportional to kW. Each time the rotor turns, it creates pulses proportional to kWh use. Those pulses are sent via a phone line to a utility computer that converts them to average kW in a defined period (e.g., 15 minutes). Demand is calculated by dividing the kWh consumed in that time period by its duration. Data is stored in the utility's central computer, but not in the meter.

A *digital* meter (also called "smart," "interval" or "IDR" for Interval Data Recording) electronically counts kWh in 15-minute (or other) intervals and divides that number by the duration of the period to yield average kW during that brief period. Data is sent to the utility via the internet, radio, powerline carrier, or other means. In a later chapter, we'll discuss ways to use smart meters to cut your electric bills.

Some digital meters store months of such data for later retrieval. Some measure it in short (e.g., 5-minute) periods which may influence how peak demand is billed in the future. A federal voluntary standard (called Green Button) calls for ¼-hourly interval metering, but (as of 2017) many utilities still measure demand in 30- or 60-minute periods. Learn more about that standard at www.greenbuttondata.org.

Smart meters enable "dynamic" electric rates based on time-of-use (TOU), real time pricing (RTP) at the wholesale grid, and other time-sensitive tariffs. They are essential for demand response (DR) programs that pay customers to reduce load when called upon to do so by a utility or grid operator.

Peak demand may be charged for the highest peak seen at any time of day during a month, or during a specified time period (e.g., 8 AM to 6 PM weekdays). The method is whatever is stated in the utility's tariff for the customer's rate class.

Despite erroneous technical literature distributed by a federal agency in the 1970s, *billed* peak demand is NOT the instantaneous electric load created by brief surges, as occurs when an electric motor is started or fluorescent lighting is turned on. Such surges last only a few seconds (or shorter) and may have no discernible impact on kWh consumption during a 15-minute interval, unless cycled very frequently. Some facility managers think they can reduce peak demand charges by turning on a building's lights one floor at a time. Once all the lights are on, however, the peak will still be reached. Staging such startups may avoid popping a circuit breaker, but it won't reduce billed peak demand.

WHY UNDERSTANDING PEAK DEMAND IS IMPORTANT

Due to changes in how power is generated and transmitted, peak demand charges may be the fastest-rising part of your electric rates. In some areas, peak demand charges now account for up to 70% of annual electric spend. Figure 3-2 shows how much (in percents) both energy and demand rates have changed in California. Each line is a different rate class. Demand charges (light lines) rose even as energy charges (dark lines) fell.

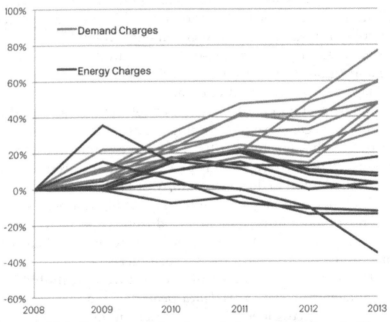

Figure 3-2. Light lines show rising demand charges for various tariffs, and dark lines show falling or stable energy charges. Demand may be the fastest rising rate component for many C&I customers. *Source: STEM, Inc.*

Peak demand charges may appear in several places in a tariff because demand may be charged separately for supply and delivery, e.g., different monthly demand charges for generation, transmission, and distribution.

There may also be a charge based on the highest annual peak demand, called a "ratchet," "capacity," or "contract demand" charge. That annual peak may be called a facility's "capacity tag." Some form of it is paid by about 60% of C&I customers. The charge may remain constant every month until peak monthly demand has remained lower for 12 consecutive months, or other defined time period. The kW being charged going forward will then be the highest level kW seen in that recent period, instead of the previous annual peak.

Some tariffs may instead (or also) levy a *minimum* ratchet charge. Imagine, for example, a commercial office building that has lost many tenants during a recession, or, through installation of an on-site cogeneration plant, significantly cuts its peak demand. If that customer's peak monthly demand drops below (e.g.) 50% of its annual peak, it may still be charged for that 50% until a year has passed. That number may be annually adjusted in the same manner as described above.

Much to the chagrin of energy managers and ESCos that promised immediate and specific dollar savings from their energy efficiency work, ratchet charges have sometimes delayed or reduced them when not properly taken into account.

Figure 3-3. Transmission towers move high-voltage power across the nation. *Photo credit: US DOE*

How To Almost Lose $120,000

Due to a prior service issue, a utility offered as restitution a temporary discount off of its delivery tariff to customers in the affected part of a city, but only to those whose annual peak demand was 400 kW or less.

An energy consultant (Tony) to a real estate manager (Nick) saw that one of Nick's buildings was in that area. Tony suggested that Nick apply for the discount.

Nick replied, "Oh, I already rejected that offer. We're too big to get it. Last month, we used 120,000 kilowatts." Tony was stunned: 120,000 kW is 120 megawatts (MW). That's the peak demand for a small city, not one mid-sized building.

Tony realized Nick may be confusing kWh with kW. He asked him, "I think you mean you used 120,000 kilowatt-*hours*, not kilowatts, right?" After a brief silence, Nick replied, "What's the difference?"

The building's peak demand was only 396 kW. It was thus eligible for a $60,000 a year discount for 2 years.

Tony fixed the situation, saving $120,000 (and maybe Nick's job) merely by filing some paperwork.

Chapter 4

Typical Components in a C&I Electric Bill

While their names may vary, some of the most common are:
- customer charge (fixed for each rate class)
- metering/billing (fixed for each rate class)
- generation (i.e., "supply")
- transmission
- distribution
- fuel adjustment charge (FAC)
- program charges (e.g., efficiency, low-income, grid, stranded costs)
- power factor (PF), also called "reactive power"
- extra utility equipment (specific to a customer)
- taxes (not found in tariff)

Let's look at each of them.

FIXED CHARGES

Every bill has a fixed "customer charge" based on rate class. Think of it as a connection fee. It appears each month, even if no electricity is used.

Some utilities also charge a separate monthly fee for each meter and/or for billing. These too are irrespective of consumption or demand.

If extra distribution equipment, such as backup feeders and/or transformers, was requested by a customer (e.g., to ensure greater reliability), it may be charged via a fixed monthly fee appearing as a separate line item on the bill. It may be an ongoing leasing fee, or an amortized purchase charge that is dropped once the utility has been fully reimbursed for providing it.

MAJOR VARIABLE CHARGES

A monthly charge for each of the following will vary based on usage. Each may contain a separate consumption charge (in $/kWh) and at least one peak demand charge (in $/kW).
- generation
- transmission
- distribution (possibly with separate high- and low-tension charges based on delivered voltage).

Pricing for each of them may also vary based on:
- customer rate class (e.g., commercial, large industrial)
- time-of-day (typically on- and off-peak periods)
- time-of-year (often groups of months or seasons, e.g., summer/winter)
- tiered blocks (e.g., first 1,000 kWh at $X/kWh, next 10,000 kWh at $Y/kWh), with price ascending or descending.

As previously described, monthly ratchet charges (in $/kW) may be fixed for a year based on the highest demand seen in the last year, and may then change if a different peak demand is seen in a later year-long period.

Other Variable Charges
- Power factor (PF): Some devices (e.g., motors, magnetic ballasts, older variable speed drives) push the voltage and amperage 60 Hz wave forms out-of-phase. Utilities must compensate by increasing generator output to provide more "reactive" power (called volt-amps-reactive, or VAR). The lower a facility's PF, the more VAR must be provided, resulting in higher cost to the customer. As more digital meters are installed, customers may begin to see a PF charge if their old meters were unable to measure it.

PF below a level (e.g., 95%) found in the tariff may be billed at a fixed $/kVAR rate, but billing methods vary widely. It may be calculated as a multiplier of peak demand, or through a separate charge. For ways that it may be tariffed, and how to improve PF, go to http://es.eaton.com/PowerFactorROI/index.html

• Fuel (or energy) adjustment charge (FAC): To supply power to retail customers, utilities buy fuel and/or power from wholesale markets. To develop the base supply rates shown in a tariff, a utility assumes prices for such commodities. If the actual price is higher, the extra cost above that base is recouped via a FAC, typically charged as $/kWh.

In Figure 4-1, we see how electric pricing (dark line, left axis) parallels that of natural gas, a common generator fuel (broken line, right axis): when one rises (or falls), so does the other. If a utility assumed a price of $4.50 per dekatherm (horizontal dashed line) to generate power at its base supply rate, a positive FAC would occur when the wholesale gas price rises above it. If lower than $4.50, a negative FAC would be charged (in essence, providing a discount).

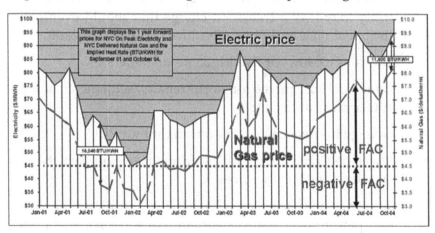

Figure 4-1. Wholesale electricity pricing parallels that of natural gas pricing. When it rises above a defined level (see dashed line), a positive Fuel Adjustment Charge (FAC) may be levied. When below that level, a negative FAC may instead be charged. *Source: NY State Public Service Commission proceedings, plus author's graphics.*

Consecutive months having positive and negative FACs have left many facility managers scratching their heads when their average electric rates fluctuated widely from month to month.

Some utilities levy a FAC monthly, while others do so seasonally or annually. About 60% of IOUs (and some PMAs) levy FACs. Those that don't may seek to recoup the cost when next their rates are being reviewed via a PUC proceeding.

PROGRAM CHARGES AND TAXES

Various small fees (adding a few percent to the total bill) may be levied as $/kWh charges. They may include:

- system benefit charge (sometimes called "universal service fee") for efficiency programs, extra costs of power from renewable power sources, and/or efforts to reduce greenhouse gasses emitted from power plants; some of those charges may instead be seen as separate line items

- in states where retail power supply has been deregulated, a utility may levy a transition charge that should eventually disappear once a utility's "stranded" costs are paid off

- subsidies for low- or fixed-income customers

- state/local surcharges for storm repairs, settlements of lawsuits against utility, etc.

- a charge to cover regulatory oversight by a PUC

- power pool/ISO charges for grid management (which may be called "ancillary services").

Taxes may be charged as $/kWh (i.e., a "utility tax") and/or as a fixed percent (e.g., called sales, and/or gross receipts taxes) of the dollar value of an electric bill. Note that sales taxes may be levied at the state, county, and municipal levels. Some customers pay almost 12% in sales taxes, making tax exemption a worthy pursuit.

MANY PIECES IN THE PIE

Consumption-based charges (i.e., based on kWh usage) are typically the biggest slice. Depending on rates and how power is used, peak demand may also be a major part of the total annual cost. See Figure 4-2 for a pie chart showing the relative impact of major rate components on a typical C&I electric bill.

Typical C&I Electric Bill

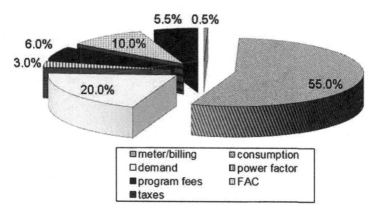

▣ meter/billing	▨ consumption
▢ demand	▥ power factor
■ program fees	▨ FAC
▨ taxes	

Figure 4-2

Chapter 5

First Steps in a Rate Analysis

Before reading a tariff, gather facility information that may be relevant to its tariff position. Assemble the following for each major electric account, and for the entire facility:

- annual consumption (kWh)

- annual and summer peak demands (kW)

- delivery voltage at the meter(s)

- kW capacity of any on-site power generation; calculate the percent of peak demand it can meet

- kW demands of major loads, especially those running during the facility's peak, e.g., lighting, space heating, DHW, HVAC, process/plug (e.g., data center, industrial equipment)

- kW and operating hours of major off-peak loads, e.g., parking lot lighting, process loads

- special conditions and tax status, e.g., religious, manufacturer, educational, medical, research

- special location: economic development zone/Industrial Development District (IDD)

Some tariffs or discounts may be related to such factors.

With that information in hand, follow these steps to analyze the rate for each major account.

1) On a bill for each account, find its rate class; look for an alphanumeric, e.g., EL 69.

2) On the bill, find the contact phone number for the utility and call it to request the name and contact information (phone and email address) for your utility account representative, unless already known.

3) Download the tariff (from either the utility or PUC web site) in word-searchable form (e.g., PDF) for that rate class. Links to sources for utility and PUC web sites may be found in appendix A.

4) Carefully read just the sections that apply to your rate class(es).

5) Isolate unclear terms and acronyms. Look for definitions in the tariff's introduction or definition sections.

6) Identify and quantify each rate component (e.g., $/kWh for generation) seen in the tariff for your rate class(es).

7) Convert any text that describes math functions (e.g., tiers, proportions) into algebraic formulae.

8) Find and read any special provisions or riders that may pertain to your rate class(es).

If any of the above is unclear or challenging, ask your account rep to cite or provide written definitions or explanations. *Developing a good working relationship with your account rep is worth your time. You're already paying for those services through your electric bill. Such a relationship may later be crucial when seeking assistance for additional services or rate reductions.*

TIPS TO MAKE ANALYSIS EASIER

The first time you try to analyze your tariff will likely give you a headache. Here are a few tips that may smooth the process.

•   Use word processing and spreadsheet programs to simplify this work rather than just making handwritten notes.

- Within the tariff for your rate class, use the word search function to find these keywords: fee, charge, contract, security, $, ¢, deposit, penalty, payment, late, option, rider, property, site, special provision, and deferral. Highlight their location and context: they may come in handy later.

- Convert rate equations and tables into Excel™ formulae, or use energy accounting software (e.g., EnergyCAP™, Metrix™) to model them.

- Read the rate slowly several times. Re-write it in common English in short sentences. Verify the accuracy of your re-write with a utility account rep or the PUC's consumer advocate's office (nearly all state PUCs have one).

BREAK DOWN DENSE LANGUAGE

Some tariffs are written in legal (not common) language, making comprehension difficult. Reduce this problem by converting relevant parts of the tariff into everyday language.

Start by replacing arcane words or phrases such as "enumerated hereon" with common synonyms/phrases, e.g., "listed below."

Eliminate all non-essential words. Focus on the operant verb(s) and convert negatives to plain language, e.g., replace "not less than" with "at least."

Chop up long sentences that are hard to follow into short, easy-to-grasp sentences. Add commas as needed to separate the phrases.

Here's an example of an overly long and complex sentence from a tariff, and how it was changed to have the same meaning with more understandable language.

"It is further understood and agreed that when the group of buildings or parts of buildings enumerated hereon are under a common ownership or leasehold for not less than a 5 year term, of public record in the name of the Customer, the electric energy supplied to such building or parts of buildings will be added, and the separate maximum demands of such buildings or parts of buildings will be added, for the purpose of determining the amount of the bill which such Customer shall receive for service."

Here's the translation, done by eliminating unnecessary words, and replacing legalese with common English:

"When the group of buildings or parts of buildings listed below have the same owner or are on the same lease, for at least 5 years under the same Customer name, then their monthly kWhs will be summed, and their monthly peak kWs will be summed, to calculate their total monthly bill."

USE UTILITY ACCOUNTING SOFTWARE

One of the best ways to model tariffs and track spending on energy is through utility accounting software such as Energy-CAP™ or Metrix™. Those (and similar) programs help enforce the level of detail needed to wring savings out of billing data. I began using the original version of EnergyCAP (called FASER) in the early 1980s and always found it helpful.

Such software looks at each bill, analyzes it many ways, and may quickly find errors and anomalies. Refunds and/or savings greater than the cost of the software (which may exceed several thousand dollars, depending on purchased features) are not uncommon. To try to save a few bucks, some large customers have tried to re-create such professional software using in-house personnel. In one case, a large university spent almost $100,000 in personnel time trying to do so. Instead, its program missed many of the tricks developed over decades by the professionals.

If making the case to buy such software may be difficult, get a taste of its capability by trying free (but stripped down) versions found online, such as Energy Watchdog™, EnergyCAP Express™, and EnergyDude™. If using one of them finds a juicy refund or savings, that may be what's needed to shake loose the money for the real thing.

If neither option is feasible, however, bills and rates may be crudely tracked and simulated with an Excel™ spreadsheet. Some error checking techniques are easily replicated, e.g., comparing bills from one month to those seen in the same month in prior years. That option may be appropriate when only a few (e.g., less than 20) accounts are involved and/or total annual spend is less than $200,000. In such cases, large billing errors are less likely.

Note that, in all cases, software is only useful if data is routinely entered each month for each account. For any program to work well, initial setup requires entering at least 2 years of monthly data by account, with consumption, demand, and other major tariff factors shown separately. Where multiple meters are involved for a single account, their readings should also be entered separately.

Doing so takes a block of time, so consider using an intern, new hire, or accounting personnel for that task. If the job is a major undertaking (e.g., entering data for 50+ accounts), consider using a consultant well versed in the software, or the software's vendor. Employees that will only do it "when time is available" may never get it fully done. The resulting gaps and inconsistencies may end up wasting time and money, and leave you chasing statistical ghosts due to entry errors.

REVIEW ALTERNATIVE RATES

A utility is not required to enroll a customer under an optimal rate, only an appropriate rate. As a result, customers may find that some of their accounts could fit under more than one rate. Rates may have sections called "riders" or "special provisions" that alter the general conditions and/or eligibility such that they could yield lower rates, either permanently or for a defined time period.

To see if your account(s) might fit under a better rate, check to see if its characteristics (e.g., annual peak demand) might fit within another tariff's limits (e.g., minimum or maximum peak kW, minimum annual consumption). For those that do, find and review riders and special provisions that may apply to them.

Word search the full tariff for descriptors that may be relevant to your facility, e.g., "job," "medical," "business," "economic," "negotiated," and synonyms for such words. Do the same with words related to possible service levels, e.g., "primary," "secondary," "transmission," "backup," "standby," "high tension," "low tension."

For example, a client used electricity to produce a product which could be stored for later sale. Instead of paying the flat year-round industrial electric rate, he found a rate option that allowed the utility to charge a much higher rate during the summer, but a lower rate for the rest of the year. The client reviewed how a shift in production hours and months might affect his operation. He found that making and storing some product during a later shift in some months, and consolidating employee vacations to allow closing of the plant for a summer month, could avoid high power pricing while giving him a lower rate for all hours at other times. While production was shut down, his sales department continued to operate by selling the stored product. Working with a utility account rep may help find such opportunities.

WHAT TARIFFS MAY NOT SHOW

Tariffs may mention, but not quantify, a variety of charges that vary based on wholesale market conditions, technical issues, or special programs (e.g., economic development). Tariffs may contain maps showing where such programs apply or links to where more information may be obtained. They may not, however, show the dollar value available to customers located within them. Do the digging needed to determine if an alternative is viable, or is a dead end. Keep a record of such investigations, including the contacts developed in that process. Program rules may later change in your favor, so it pays to know how to quickly follow up on them. Following are charges that may be listed but unquantified in a tariff.

- Fuel adjustment charge (FAC): it may vary based on wholesale fuel and power markets, which cannot be predicted in advance. It may be a $/kWh charge shown separately on the bill.

- Line losses: Also called "unaccounted for energy" (UFE), it covers losses (as a percent of kWh volume) during the delivery process due to electric resistance, transformer inefficiencies, and theft of service. It may be announced monthly or annually in a utility web posting or newsletter. It's typically built into utility supply rates, but (in a deregulated state) it may be charged separately to a third-party power supplier, who then charges it as a line item on a customer's invoice.

- Additional distribution: Specific to each customer, it charges for extra feeders, distribution transformers, or other equipment requested by the customer. It may be a fixed monthly charge based on a lease or as an amortized purchase price.

- Program charges: They may be bundled together or appear as surcharges based on $/kWh or as a percent of the total bill. Among them may be costs for: energy efficiency incentives, utility losses due to such programs (which may be called a "revenue decoupling mechanism," or RDM), low-income subsidies, and/or purchase of a defined portion of a utility's energy from renewable sources, commonly called a "Renewable Portfolio Standard" (RPS).

- Economic development and business incentives rates may be found as riders, or referenced as rate discounts provided by a government agency. The tariff may provide a link for further information from that agency and/or maps or text defining the geographical areas to which they pertain. If a facility is adjacent to, but not within, that space, a discussion with the agency's administrator (to try to stretch the map a bit) may be worthwhile.

- "Special," "experimental," or "pilot" rates may be listed but not quantified. They may be limited to a single or a few large customers, offered as a way to keep them in a utility's territory. Such deals may be special provisions or riders very narrowly written to fit only those customers. A "special" tariff may offer an option for a negotiated rate that may not later be publicly posted.

- Off-tariff rates: When a large customer (e.g., data center, or manufacturing plant) with a large load (e.g., several percent of a utility's peak) is planning to locate in a utility's territory, it may negotiate a separate utility power contract that acts as its proprietary tariff. Such deals won't be found in the published tariff, but may be found in a PUC rate proceeding if its approval was required.

Taxes are not shown in tariffs. While collected by the utility, they are then paid to governmental entities and may vary across time and geographical boundaries. Details on them will be found in tax laws and regulations, not in the tariff.

Bottom line: to assess the spectrum of options available to a customer may require research outside a tariff. Examining several past PUC rate proceedings (posted at its web site or otherwise available) may be very educational to understanding how others have pursued opportunities, and which have been successful.

Section II:
Options for Rate and Cost Reductions

Chapter 6

(Mostly) Administrative Options

Many involve no changes to building systems (e.g., HVAC). Several (e.g., tax reduction) exist outside tariffs, but may be related to the type of electric load (e.g., product manufacturing) or how bills are rendered and paid. Each may help cut costs related to electric bills.

Note: Few or none of these options may work at your site. Some may save only a fraction of a percent off the total bill and/or its handling costs. Successfully pursuing several may, however, add up to a lucrative use of your time, with little (or no) out-of-pocket expense.

Where an option is not presently available, consider raising it during a rate proceeding (called "intervention" and discussed in Chapter 11) so that it may be created, e.g., by further unbundling rates or adding a new option.

AVOID, SHIFT, AND/OR REDUCE TAXES

Find and utilize (and maybe stretch) exemptions already existing in your state and local tax codes. Customers that can, for example, support a claim of being religious or non-profit may avoid sales and other types of taxes. In a major city in the northeast, an old church to which land was bequeathed over more than a century was able to have many of the buildings that were later erected on those grounds to be considered "religious property." It pays no sales tax (which would otherwise exceed 8%) on the power feeding those buildings, which contain millions of commercial square feet.

Start by meeting with your local economic development agencies. Many cities, counties, and some states have them. Their charter includes helping businesses to minimize taxes to keep jobs in their jurisdictions. If no such agency exists, contact a local business organization (e.g., Chamber of Commerce) or an association related to your firm's industry (e.g., a hospital or manufacturing association) for assistance. If other customers performing the same function have secured an exemption, then so might your firm.

In many states, for example, power used to manufacture goods may be exempted from sales tax if that tax will later be collected at retail for the sale of those goods. Where a commercial real estate landlord sub-meters its tenants, it may avoid sales tax on the power it purchases for its tenants if the sales tax is later charged to the tenants on their sub-metered bills. The landlord then collects that tax through tenant bills and sends it to the taxing authorities.

If too challenging, outsource this task to a company that specializes in securing tax exemptions (many are accounting firms). Some will, for a flat fee or a percent of savings (instead of an hourly rate), pursue the exemptions for you.

Trimming Off The Sales Tax

A large food packaging facility had many tenants, each of which used the facility's refrigerated storage to hold, dress, and package meats for later sale to retail stores and restaurants. Electricity was a significant part of the facility's operating costs. Sales tax (state and city) on the electricity exceeded 8% of the total bill.

At the urging of the facility's energy consultant, the managers secured the services of an accounting firm well versed in tax law. A provision of the sales tax law forbids multiple taxation on the sale of a product: under the law, sales tax could be charged only once, and only at the final point of sale. Because the meat was a product whose manufacture included use of electricity to cut, package, and freeze it, that energy was essentially part of the final product. If that electricity was sales taxed, and the product was again later sales taxed (e.g., by a restaurant), double sales tax would be levied.

Demonstrating how much of the facility's electricity went toward production (as versus office lighting and other non-production loads) involved a one-time survey and analysis, as specified by the tax law. The sales tax exemption was secured, saving many thousands of dollars a year. The payback period on this "paper pushing" was measured in weeks, not years.

MINIMIZE BILL PROCESSING COSTS

When many accounts are involved, merely managing electric bills may be a significant expense. Following are four ways to reduce that cost, each of which may involve your utility. To determine which are available (or could be secured), start by asking your utility account rep.

EDI: Handling each electric bill (i.e., opening, logging, printing a check, mailing it) typically costs $5-$10 in accounting personnel time. Cut such costs by asking for utility bills to be provided in electronic data interchange (EDI) format. That process formats the billing data so it may go directly into a customer's accounts payable (AP) program. Doing so avoids paper bills, speeds up accounting, and essentially eliminates entry errors. This option may be most appropriate for customers having many (e.g., 50+) accounts.

If EDI is not presently offered, work through a business or professional association to request that it be implemented by the utility on behalf of its members, or push for it during a PUC rate proceeding. For details on EDI, watch this short video: www.energycap.com/products/energycap-enterprise-videos/thank-you-receiving-bills-electronically?autoPlay=true.

Summary billing: Some utilities and third-party power suppliers will, upon request of a large customer or group, provide a single bill that shows all data by account number and a monthly total, allowing payment using a single check or transaction for all accounts each month.

Pay early via EFT: While utilities typically require payment within 30 days of a bill's date, a few (in the past) provided a small discount or incentive for large customers if payment was made bank-to-bank via electronic fund transfer (EFT) in 10 days or less. During times of high inflation, doing so gave the utilities the "float" on the diminishing value of the currency. Many large customers now use EFT as the default method for paying large monthly bills since doing so is more secure and manageable. A discount for early payment via EFT may again become available if inflation rates rise significantly (e.g., over 7%). For readers too young to have lived through high rates: in 2017, the prime rate was only 3.75%, but it was ~11% in 1974, ~22% in 1982, and remained above 9% from 1982 through 2001.

Bill paying services: For very large firms (e.g., chain stores, banks), bill paying services (e.g., Ecova) receive and pay their utility bills, charging a small percent of their value for the service. They also provide summaries and "sanity" checks that may find errors and overpayments. Speak to your accountant regarding them. This process may require that your utility send duplicate bills to the service, so a chat with your utility account rep may be needed.

AGGREGATE METERS/ACCOUNTS

Some tariffs allow multiple accounts held by a customer on the same property to be gathered onto one account. Some utilities call this "conjunctive" or "coincident" metering. Doing so may reduce several costs:

- a single large account's connection charge may be less than the sum of many smaller connection charges for multiple utility accounts held by the same customer

- billing charges (for both the utility and internal to the customer) may be lower

- if the tariff has a *descending* block rate, the combined consumption may shift much of the usage by the now-aggregated account into a cheaper incremental rate

- peak demand seen on the single new meter is almost always lower (it can never be higher) than the arithmetic sum of the non-coincident peaks seen on the separate demand meters previously measuring the small loads. Where demand charges are high (e.g., 20% or more of the total bill), savings from *coincident* peak demand reduction alone may make this process worthy of consideration.

Note that the definitions of "customer," "property," and "site," (as seen in the tariff and local tax laws) may be key to this process. Remember how I told you to highlight them when you found them in the tariff?

Getting The Doctors Together

To handle expansion of its facility, a hospital needed new offices for its many doctors, other personnel, and small clinics. To do so, it bought an adjacent apartment building and converted the apartments into medical offices. When the building was purchased, each of its 100 apartments had its own utility meter and account. In its haste to move personnel into the building, the hospital simply took over those small accounts, all of which were on a residential rate.

The hospital's energy consultant calculated that the total cost to serve those 100 accounts was higher than if they were combined onto one account. After much persuasion, the hospital applied for a rate change that resulted in removal of the residential meters and installation of a single new utility meter for the building at a cost of about $40,000. Initial calculations indicated that the job would have about a 4-year payback.

After a year of operation, total savings were higher than expected because:

- the varying schedules of the doctors and clinics resulted in a lower peak demand than projected

- most of the kWh used fell into a cheaper descending block rate
- the now-aggregated load was sufficiently high to secure a better third-party supply rate than expected (the facility was in a state allowing competitive purchasing of supply).

In the end, the simple payback was just over 2 years. No energy efficiency upgrades were involved, and no changes to occupant behavior were required. If the hospital wants to later install sub-meters for each new office, the prior meter slots (called "pans," each wired to a separate apartment) remained available for such installation.

RE-CLASSIFY LOADS/ACCOUNTS

Depending on its characteristics, an account may fit under several different rates. Some creativity may be needed to make the case, but the end results may be well worth it. Here are a few examples.

- A seminary is essentially a religious college, with all the trappings of an educational institution, including lots of computers. Its energy consultant noted that, as it grew, the seminary added buildings (and accounts) with significant loads. As each was opened, the utility placed it on the same "religious" rate as the institution's other sites.

One of those new buildings was essentially a data center, with a large computer bank running 24/7 as its primary load. Based on its peak kW demand, that site could also qualify as an "industrial" customer. An analysis of its electric bills showed savings if switched to that rate. No physical change— not even in the meter—was necessary. Only a request for a "change of service" with a small one-time fee was involved, yielding a payback measured in weeks.

- A video game developer created software, burned it onto DVDs, printed its labels, and (via automation) packaged the discs for sale to retail stores. When first opened, the utility (not being asked otherwise) placed the customer on a standard "commercial" rate. By playing up the production and packaging processes and its nearly flat load profile, its energy consultant persuaded the utility to switch it to a slightly lower "manufacturing" rate. That, in turn, was sufficient to secure a sales tax exemption for a large portion of its electricity consumption.

- Health clubs are typically large consumers of domestic hot water (DHW) for showers, hot tubs, and pools. While a gas-fired boiler would have been preferable, one instead had an electric boiler with a hefty peak kW load. That boiler fed multiple DHW storage tanks serving various parts of the gym. The club's energy consultant found a "DHW" electric rate that was cheaper than the rate on the account serving the building. By opening a new electric account just for the electric boiler (which required a separate panel and wiring), that load was switched to the lower DHW rate, cutting the total electric cost for the facility. Once again, an acceptable payback period was secured without changing the facility's process equipment or operations.

- Marijuana growers in California became eligible to switch to a utility's agricultural rate if they met several conditions (e.g., at least 70% of electricity is used to grow plants). Grow lights, fans, etc. are a significant part of the cost of that process.

Some tariffs have defined customer types by their Standard Industrial Classifications (SIC), a system later replaced by the North American Industrial Classification System (NAICS); see the glossary for details. Due to how they view their operations for tax purposes, some energy customers may fit under several different codes. Where an industrial tariff rate may offer savings over a customer's existing rate, there may be value in adjusting the customer's SIC code to allow it to shift to that rate. Care is needed, however, since doing so may have federal tax implications. Consult with the firm's accountant before proposing such a change.

LOWER SERVICE FIRMNESS

Standard utility service is "full requirements," meaning that the customer will get as much power as he wants (within the capability of utility feeders and transformers), whenever the customer wants it, at a regulated price.

But some utilities offer a reduced rate or an incentive if they are allowed to interrupt part or all of a facility's electric service when needed to avoid an outage or very high wholesale power price. Under an "interruptible" rate, a customer may agree to cut back (e.g.) 30% of his load and always pay less for that portion of his power, whether or not the utility calls for an interruption. If the customer fails to cut load, power is still delivered and the lights stay on, but there may be a high penalty for the power that is used during the interruption's hours.

For facilities with flexible loads and/or on-site backup generation that may be run when requested by a utility, such a rate may offer significant savings or revenues. Some utilities call such options "load management" or "demand response" rates or programs. Here are a few examples.

- A metal processing plant ran three 8-hour shifts. Its preferred schedule involved annealing its feedstock (i.e., heating at high temperature to soften it and reduce brittleness) using a giant electric oven. To secure a rate cut even when not running the oven, it allows the utility to request that process be run during a later shift up to 4 times a year. Such requests typically came only once or twice a year (or never), with sufficient notice to allow a manageable switch involving only a few hours of overtime labor. Net savings were essentially free to the plant.

- When a backup or emergency generator is available that may be run without violating emissions regulations, a utility may offer a slightly different interruptible rate. In order to satisfy those rules with its emergency generator, a college installed a nitrous oxide (NO_x) reduction system on the unit. Like the metal processing plant, it was then able to reduce its apparent meter load whenever the generator ran at the utility's request. Doing so qualified a load equivalent to the generator's output as "interruptible," securing a lower rate for that load whether or not the generator ever ran.

- By upgrading its energy management system (EMS) so it could reduce (not shift) load when called upon by the utility or grid operator, a shopping mall was able to participate in several demand response (DR) programs. One program credited the mall's electric bill for measured load reductions (with no penalty for failure), while another paid for the right to request a load reduction on short notice (with a penalty for failure). DR is discussed in more detail in Chapters 8 and 9.

SEEK A CHP DEFERRAL OPTION

Savings may be available merely by threatening to install on-site power (such as a microturbine or diesel generator, see Figures 6-1 and 6-2). Some utilities have existing cogeneration deferral rates that pay a customer *not* to do so, making this option relatively straightforward.

If your utility is still regulated (or owns generation), hire an electrical engineering firm or en-

Figure 6-1. Microturbines are a source of on-site power. *Credit: US DOE*

Figure 6-2. Diesel generators are often used in cogeneration systems that provide both on-site power and heat. *Credit: US DOE*

ergy service company (ESCo) to plan out a viable CHP system. Use the plan as a basis to seek a CHP deferral rate that keeps you as a full utility customer for (e.g.) the next decade. Accept that discount *instead* of installing the system. As with other options discussed below (e.g., transmission bypass), a utility is unlikely to take you seriously without seeing a valid technical analysis showing that an option could cost it revenue while being cost-effective at a customer's site.

FIND BILLING ERRORS USING A
PROFESSIONAL UTILITY BILL AUDITOR

By some estimates, nearly 2% of utility bills contain at least one error. In many cases, utilities eventually correct their errors by providing billing "adjustments" without outside intervention. But sometimes mistakes are not caught, and a customer ends up being overcharged. Regardless of the injustice involved, a customer may have only a limited time (e.g., 2 years) to seek a refund.

Utility bill auditors are consultants that review bills and rates to help customers recover refunds and/or find lower rates. Many large customers use an auditor under no-fee contracts wherein the auditor is paid a proportion (e.g., 25%) of what it recovers for the customer.

Such arrangements are attractive because they look like no-risk deals, but be sure to read and fully understand the contract before signing. There are many honest auditors, but a few work on the fringes of professional acceptability. Some customers have felt "burned" when they had to (for years) split savings with an auditor just to correct a mistake that should have instead been caught by an attentive employee.

If a billing issue is due to a recurring error (e.g., a wrong rate) causing a monthly overcharge, the auditor's contract may entitle it to receive a percent of the monthly savings for several years (typically 3 to 5), yielding a bonanza for the auditor. Had it not been caught, however, the error may otherwise have continued for years (or forever), so this arrangement is considered a fair way to pay for the service. Because the vast majority of bills have no errors, that fee covers the auditor's time investment while examining many bills that produce no savings.

When an account is opened with a utility (by requesting power service from it), the utility will choose the rate, unless a customer requests a different one. Some auditors profit merely by switching customers to better rates, or finding riders that provide a discount off a standard rate, e.g., for incenting a customer to remain in an economic development zone. In states that have deregulated their IOUs, many auditors act as brokers to seek lower supply rates (see Section IV: Competitive Retail Power Procurement) and claim part of the resulting savings.

An auditing contract may create an exclusivity lasting years, and may limit customer use of any other auditing services during that time. To be sure you're getting the most out of such a commitment, review the auditor's tariff knowledge and refund experience, check two client references, and do a Google search with the auditor's name and "scam" or "complaint" in the search field.

Any proposed bill auditing agreement is negotiable. Adjust a proposed contract so it specifically limits the auditor from making claims resulting from a customer's own efforts (based on documentation). Bar the auditor from claiming a fee for the following:

- efficiency upgrades pursued by the client on his own. An auditor may claim to have made the customer aware of the opportunity prior to the customer initiating the upgrade, thus claiming responsibility for it.

- when a customer enrolls in an incentive program based on its location (e.g., demand response, or an economic development program). If the auditor can show (e.g., via an email) that it had informed the customer of a program before it enrolled, the auditor may try to claim a piece of the savings, even if the customer had learned of the opportunity completely on its own at an earlier time.

If the auditor refuses to accept such terms, find a different auditor. A list of web addresses for many auditors, and a suggested scope of work for the auditor, may be found in appendix B.

UTILIZE FINANCIAL TECHNIQUES

Several tricks that utilize a customer's financial and/or energy procurement acumen are worth consideration.

- If a facility has purchased extra utility distribution (e.g., spare feeders for resiliency) and is paying it off via a line item on its electric bill, take advantage of market interest rates (if lower than those given by the utility) by re-financing that cost through a loan, bond, etc., that buys out the remaining balance at a lower total cost.

- A trend among utilities is to push more (if not all) of their supply costs and cost risk onto their customers using fuel adjustment charges (FAC) or hourly power pricing. While many utilities control price volatility by hedging for their

How Not to Dress for Success

One of my clients was a rapidly expanding clothing line that at the time was opening (on average) a new store somewhere in the U.S. every day. Rather than reviewing the tariffs in effect wherein the new stores were located, those responsible for setting up new utility accounts simply allowed each utility to place them on standard commercial rates.

Not having anyone in the firm versed in utility costs, the client had a contract with an engineering consultant whose services included securing savings by correcting utility billing errors. The arrangement was fairly common: a consultant finds an error, and gets paid a percent of the resulting saving, in this case, 50% for 3 years. Such savings seem like "found" money because the utility overcharge might otherwise have continued indefinitely.

Upon reviewing how much was being paid for that service, it became obvious that a better way to pay the engineering consultant would be via a flat fee per account for initially enrolling an account at the best available rate. That could be done for only a fraction of what would otherwise have been paid to secure it later.

The client thought that was brilliant, and proposed the flat fee idea to the engineering consultant. The latter became enraged, and threatened to sue my client for "loss of expected revenue" based on a continuation of the enrollment errors it expected to correct.

In the end, the client instead chose to let its contract with the engineering consultant run out and pay the 50% fees until their terms ended. It later hired someone who understood tariffs to open its new utility accounts.

residential customers, they may not do so for C&I customers. Customers experienced with financial derivatives may use that in-house capability to "hedge" their own FAC. Where a FAC is a routinely significant part of utility's supply charge (e.g., is a full pass-through of wholesale market pricing), a customer may buy financial options on the main fuel that the utility uses to generate power (e.g., natural gas). If that fuel's wholesale price rises, so may the value of the options. They may then be sold at a profit to help pay off a high FAC. The utility is not involved in any way. Dealing with such derivatives is only for those having experience with them, and an ability to take some financial risk (e.g., if gas prices tumble, their financial options may become worthless).

- If a muni or co-op utility has gas-fired generation, and a large facility buys its own natural gas via a non-utility supply contract, it may be feasible to negotiate a "tolling" agreement with the utility. Under it, the facility buys extra gas and directs it (through its supplier) to the utility's power plants where it is burned to supply the facility's electricity. Doing so may allow elimination of the customer's FAC and the base fuel cost built into its electric supply rate.

Tolling may be feasible where a utility does not already hedge its gas purchasing (i.e., buying it in advance at a known price) for its C&I customers. For a utility to accept a tolling proposal, a customer should already be a large gas buyer (e.g., millions of therms a year), have a good working relationship with a gas supplier, and be among a utility's largest electricity users.

To see if your company is among the top 20 behind your utility, review its Form 566 filing at https://elibrary.ferc.gov/idmws/search/fercgensearch.asp. Choose a date range of 1 year, and check off "Electricity" in the "Library" options. In "Class/Type Info," choose "Form 566" and, In "Text Search," enter your utility's name.

Chapter 7

Utilizing Metering and Dynamic Rate Options

Submetering tenants will make them fully accountable for their consumption.

In many large facilities, a single department (e.g., Facilities Management or Purchasing) is responsible for paying electric bills incurred by all tenants or divisions (e.g., dorms, labs, administration). Making each responsible for its usage may cut the entire facility's bill and perhaps its average electric rate.

First, determine how those tenants or divisions are billed for their electricity, or if that cost is instead built into their rent or departmental budget. The latter method is essentially a ticket to waste energy: if, for example, lights or A/C are left on at night, it will have no effect on their expenses.

To create accountability, a facility may divvy up its monthly electric bill based on each department's (or tenant's) floor area as a percent of the total, and send each an allocated monthly bill based on it. Unfortunately, such a process tacitly assumes that (e.g.) a laboratory uses the same kWh per square foot per year as an office or dormitory. In reality, the former may instead use 5 to 10 times that of the latter.

Alternatively, some facilities use connected electric load (as a percent of the total) developed by a one-time load survey. This too may prove unfair when a tenant later installs a 24/7 data center as part of its office, with all others then paying most of the extra cost it creates.

Here's a useful rule of thumb: when a tenant or department chief starts receiving monthly bills based on actual metered usage, that person tends to become more energy conscious, shutting off non-essential loads whenever possible (especially at night). Monthly kWh may drop by 15-30%, saving money for both the department and the whole company.

Figure 7-1. Sub-metering tenants and facility departments may offer a way to reduce usage merely by how it is metered. *Photo credit: US DOE*

The ideal way to measure a tenant's or department's usage is through a sub-meter. In the past, installing sub-metering was sometimes complicated and expensive due to the costs of access and wiring, but wireless and web-based metering systems have made this option more viable. Services are available that will install sub-meters at little or no cost under a long-term (e.g., 7-year) lease that may include meter reading and billing services. That process may be competitively bid, and the extra cost for those services included in the $/kWh rate charged to the tenant. In the end, the value of the reduced usage may exceed the leasing cost for the sub-meters, resulting in a net savings to the facility.

Note that limits may exist in a tariff (or state law) regarding the percent markup that a landlord (or his sub-metering service) may charge tenants for their electricity. Some commercial landlords profit from that markup and, as a result, have resisted energy efficiency projects that could reduce its value. If you are a sub-metered tenant, verify that your electric bills do not show a markup greater than allowed by the tariff or law. Surprisingly, some states lacked appropriate regulations until commercial tenants raised a stink about the issue. Ohio didn't fix the situation until 2015.

For landlords with commercial tenants, a better solution is to have each set up its own account with the utility so sub-metering is unnecessary. The entire burden is then shifted to the end user. Doing so, however, may entail alterations of a facility's internal wiring. To avoid that cost, new or renovated facilities that expect to have commercial tenants should be wired to allow eventual direct utility metering.

In some types of facilities (e.g., malls or shopping centers), a hybrid approach may be appropriate. If the landlord presently pays for common area loads (e.g., parking lot lighting) that are mixed into the facility's single bill, it may make sense to segregate those loads onto a separate electric account with its own utility meter. If that load runs primarily during off-peak hours, a time-of-use (TOU) rate may be chosen that yields a lower monthly cost for it.

The end result may be a rise in the tenants' average rate and a savings for the landlord. In essence, the sequestration of the parking lot lighting load onto a better rate terminates a prior subsidy hidden in the tenants' electric bills.

Before making such a change, the monthly cost under the TOU rate should be carefully modeled. Care is needed to ensure that, under no circumstances, will the parking lot lights be turned on (e.g., to find burned out lamps) during the on-peak hours, lest savings then be lost. One-time costs may also be involved: the utility may require installation of a separate electric panel and may levy a one-time charge for installing its new meter.

HOW UTILITY METERING IS CHANGING

A revolution in electricity metering is underway that is opening doors to new ways to manage power usage. How you are metered for electricity may offer several ways to cut electric bills.

Early in the 21st century, electric metering began to change. For most of the 20th century, metering depended on manual monthly readings, with the difference between readings showing usage during a month. That method is disappearing as digital meters (also called "smart" or "interval" meters) replace mechanical units.

At the beginning of 2016, the DOE's Energy Information Administration (EIA) found that 58.5 million smart meters were operational in the U.S. out of a total of 144.3 million meters, yielding a 40.6% penetration rate. Rapid deployment was spurred by the American Recovery and Reinvestment Act of 2009 (i.e., the "stimulus" to reverse the 2008-12 recession). It provided billions of dollars to utilities that installed millions of meters between 2009 and 2014. Since then, the pace has slowed and become uneven across the country. Some utilities are changing all their meters, while others are doing little or nothing. Ask your account rep for the status of that activity in your area.

Where interval meters have been installed or are planned in the near future, customers may find a variety of ways to use the meter data to minimize their electric bills. As previously described, smart meters measure usage in short intervals, e.g., 15, 30, or 60 minutes (maybe even shorter in the future). Using charting and other data tools, customers (or their consultants) may analyze usage across time in much greater depth than possible with monthly billing data.

A BRIEF TUTORIAL ON INTERVAL DATA

Smart meters produce interval data (see sample in Figure 7-2). A chart of a day's kWh usage in 15-minute intervals is called a "load profile" or "load shape." The average kW demand in each interval is merely the kWh in that time divided by ¼ of an hour (i.e., 15 minutes). Figure 7-3 is a typical profile for an office building: low kW demand in late evening and early morning hours, and a peak in late afternoon. A load shape shows details and trends in building operations and behavior that may not be seen in monthly billing data, helping many customers find ways to cut their costs Following are a few examples of how such charting quickly revealed problems needing attention. For an in-depth discussion on load profiling using interval data, see Appendix C.

Date and Time (EST)		Use (kWh)
1/1/2011	15	650
1/1/2011	30	430
1/1/2011	45	310
1/1/2011	100	270
1/1/2011	115	220
1/1/2011	130	310
1/1/2011	145	340
1/1/2011	200	620
1/1/2011	215	1330
1/1/2011	230	1250
1/1/2011	245	1070
1/1/2011	300	910
1/1/2011	315	730
1/1/2011	330	790
1/1/2011	345	760
1/1/2011	400	3860

Figure 7-2. Interval data show usage in 15-minute increments by date and time. *Source: author*

Find Electricity Wasted at Night

In load profile chart (Figure 7-3), what's going on in that building to cause a 500 kW load at 4 AM? In many cases, it is equipment left running that facility managers assumed was turned off, perhaps by an energy management system (EMS) that has failed. Tracking its interval data showed a large chain store that its off-hours lighting control system was broken, leaving stores lit all night. A quick fix resolved the problem.

A good exercise for a new hire or intern is to have that person walk the building in the early AM hours and log everything that is still on, preferably with its approximate wattage or horsepower rating. The goal is to account for most of the night time load seen in the load profile, and then find ways to better control its contributors. A common finding is hydronic heating

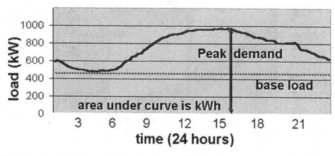

Figure 7-3. A standard weekday load profile for a commercial office building. The load peaks in the afternoon at about 1,000 kW but does not drop below about 450 kW even in the middle of the night. *Credit: author*

pumps left running during the summer when no space heating is needed. Even if no EMS exists, addition of a relay actuated by an outdoor temperature sensor could shut off the pumps whenever outdoor temperature exceeded 60°F.

Cut Peak Demand through Better Planning

Electric fork lifts in a food warehouse were being re-charged each day around 3 PM (when the fork lifters' shift ended), adding almost 54 kW to the afternoon peak. A low-cost timer and relay were added that instead delayed the charging until 10 PM. Before the morning shift arrived, the fork lifts were fully charged, peak demand had been cut, and the charging done with cheaper off-peak power. Some chargers have timers built into them that may do the job at no additional cost.

A college football stadium boasted 180 kW of field lighting. An examination of its billing data found that the site's peak de-

Figure 7-4. The charging load of electric fork lifts may be timed to reduce peak demand and take advantage of time-of-use (TOU) electric rates. *Photo credit: author*

mand was so high that it accounted for ~95% of the monthly bill. The site was rarely used during weekdays but—every now and then—peak demand would spike in the late afternoon. The utility sent me a letter saying it was charging the highest legal average electric rate, but would have charged me even more if regulations had allowed. I immediately dispatched one of my minions to investigate.

She found the stadium empty and initially saw nothing wrong. Suddenly at 3 PM, while it was still bright outside, all the field lights were turned on! She saw that the custodian had done so, and asked why. "Well, I've been told," he said, "that the coach didn't want to see any burned-out lights during a night game. So, once a month before my shift ends, I turn 'em all on to see which might need to be replaced." We eventually cut a deal wherein he would come in on a Saturday and test them when (at the time) demand charges were much lower. We agreed to pay him time-and-a-half for his overtime, which was far cheaper than the weekday demand charge.

Identify Loads that May Be Temporarily Reduced

Some loads may be briefly reduced to cut peak kW charges and/or participate in demand response (DR) programs. At a laboratory, rooftop air handlers having variable speed drives (VSD) were automatically sequenced so that—when a peak was being approached or a DR call came from the utility—a few briefly ran at a lower speed, after which they would return to normal and others would reduce their speed. Because of the approximately cubic relationship between motor speed and kW load, a 20% reduction in motor speed and air flow (not noticed for 15-minute periods) yields almost a 50% drop in fan motor kW (80% x 80% x 80% = 51.2%). The end result was a continually lowered aggregate peak load for the air handlers.

DR programs require interval meters to show the utility or grid operator exactly when and how much (relative to a baseline) a customer cut back on its usage. Such programs may cost a customer nothing to participate, and are further discussed later.

Choose an Alternative Electric Rate

Depending on the type of customer and when it uses power, annual cost may be lower under a time-of-use (TOU) rate, if available in the tariff. Smart meters are necessary for such rates. By reviewing past interval data, a customer may model what his cost would have been under a TOU rate, and then request a rate change.

A small church on a standard religious rate closed during the summer (it had no air conditioning), and was used at other times primarily on weekends and evenings. Based on that schedule, a rough load profile was simulated. Were it to shift to the available voluntary TOU rate, its annual cost for electricity could be reduced by almost 10%. The utility installed an interval meter for that purpose at no charge, with the only caveat being that the church had to remain on the rate for at least a year, or else pay a penalty to switch back sooner.

HOW TO GET THE DATA

If your utility has already installed a digital meter, its interval data may be available online or by email from the utility at low or no cost. If a mechanical meter is still in use, the tariff may spell out how to have it replaced by a smart unit (at a one-time charge by the utility), allowing direct access to its data output. Before doing so, ask your utility account rep if your meter is already slated to be replaced with a digital unit, and if that schedule may be expedited.

If a smart meter is not on the horizon, several ways exist to generate your own interval data without it.

1.   Inexpensive (~$250) strap-on meter reading devices pull interval data from meters without involving the utility. Two vendors of such devices are:
      • BlueLine Innovations (www.bluelineinnovations.com)
      • WattVision (www.wattvision.com). See Figure 7-6.

Since they attach to the outside of meters (which are utility property), it's a good idea to clear the attachment with a utility account rep before purchasing and installing them.

For the oldest mechanical meters, such devices may use an infrared laser emitter mounted on a meter's glass cover. It sends out a beam that measures the speed of the meter's rotating disk (common to most mechanical meters) which is proportional to kW. One device sends the data to a local PC via WiFi and a USB receiver, or other communications methods. Data from another may be accessed via a secure web site for a few dollars a month.

2. Some C&I customers still have pulse meters that send data through telephone lines to a utility computer. With utility permission and cooperation, such meters may be opened, allowing a connection of a pulse reader to the meter's KYZ ports. Often called "shadow" meters, such readers then send a duplicate stream of that data in real time to a customer's EMS, or to a data collection system that translates it into standard 15-minute (or shorter) interval data readings. To find such equipment, Google on "electric meter pulse reader." Some are wireless, sending their data directly to your computer via WiFi.

3. Data loggers use clamp-on current transformers (CTs) to measure and record current in a feeder or circuit, acting like temporary meters, without cutting into wiring or interrupting power flow. Many good and inexpensive data loggers are available, among them:

   • www.dentinstruments.com

   • www.onsetcomp.com.

Figure 7-5. Dent Instruments data logger acts as a temporary meter using current transformers. *Photo credit: Dent Instruments, Inc.*

Where a utility has installed smart meters, it may not yet have started to read their interval data, or make it available to customers. Nevertheless, those meters may still be outputting infrared pulses through a port on the front or atop the meter (see clear nub near the center of Figure 7-7). Those pulses may be read by other versions of the strap-on devices mentioned above (at about the same price). Some can read intervals as short as 7 seconds (though doing so generates an enormous amount of data).

The Association of Energy Engineers (AEE) offers a professional webinar on charting and analyzing interval data. See www.aeeprograms.com/realtime/LPonline/for details.

Figure 7-6. WattVision strap-on device reads data from a digital meter and sends it to the customer. *Photo credit: WattVision*

EVALUATING A TOU RATE

Interval data may be used to model the cost of a facility's electricity under a time-of-use (TOU) rate. Where such rates exist, the utility may already possess sufficient data to determine the potential savings (or loss) had a

Figure 7-7. Clear vertical nub near center of photo emits interval data in the form of infrared pulses that may be monitored by economical strap-on devices. *Photo credit: author*

customer been on that rate for the prior year. Ask an account rep if such an analysis may be done at no, or a low, charge by the utility. If not possible, a customer (or its consultant) may request or collect a year of data and, using a spreadsheet, do that analysis based on the utility's TOU tariff.

TOU rates are generally much higher during weekday days than on nights and weekends (and maybe holidays, if so stated in the tariff). Larger differences may be seen between summer and non-summer weekday days.

In Figure 7-8, price is on the y-axis and months of the year are on the x-axis. TOU pricing (top line) is double the standard rate (middle line) during the summer months (inside the shaded box) but, at night, weekends, and holidays (bottom line), it's about half.

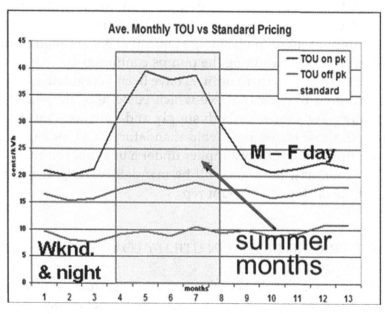

Figure 7-8. Time-of-use (TOU) pricing across a year. The top line is on-peak, the middle line is the standard electric rate, and the bottom line is off-peak pricing. Note that on-peak pricing is nearly double standard pricing in the summer months, but off-peak pricing in those (and all other) months is about half the standard rate. *Credit: author*

With a TOU meter, savings opportunities may arise unrelated to the TOU rate. Data from a TOU meter may show periods of high kWh consumption when it should instead be low (e.g., nights, weekends), due to unauthorized facility use or failed controls. Correcting the problem may cost little or nothing.

Cost-cutting options under a TOU rate may exist by time-shifting loads and minimizing on-peak kWh consumption. A simple example is electric dehumidification. Essentially A/C units controlled by a humidistat instead of a thermostat, some facilities run them 24/7 in basement storage areas to limit humidity that may otherwise support mold growth or odors. In some cases, it may be possible to allow humidity to vary slightly more by having them instead run only during each day's off-peak period, controlled by an inexpensive plug-in timer. As long as the units run sufficiently during the night, spaces may remain dry enough while reducing daytime peak demand running on cheaper off-peak power. Agricultural facilities that pump water may run their pumps at night to fill storage tanks that are emptied during the day rather than running the pumps continuously.

Note: in states where utilities have been deregulated, a peak kW threshold may exist above which customers are placed on mandatory TOU rates for both supply and delivery. Power customers in those states may avoid mandatory TOU for *supply* by buying from a non-utility supplier under a bilateral contract. The *delivery* rate may, however, still be mandatory TOU. If unsure, consult the tariff or an account rep.

POWER PRICING BASED ON UTILITY LOAD PROFILES

Utility rates for classes of customers (e.g., small commercial) are based on *averages* of many load profiles developed during utility samplings within each rate class. Some facilities with unusual equipment (such as thermal storage) or long operating schedules (e.g., hospital, multi-shift industrial plant) may, however, have flatter load profiles. As a result, using a standard utility load profile to develop competitive supply pricing (where available) for such facilities could yield a higher average supply rate. Without hourly metering data, a competitive supplier may have no choice but to use standard profiles to develop its pricing, which may then not be much better than that of the utility. Such standard competitive supplier rates may be called "rack rates" because they are essentially "off-the-rack" products, not tailored to a particular customer's profile.

A SLIGHTLY EXTREME EXAMPLE

Let's compare two 300-kW loads having similar load factors, i.e., average demand divided by peak demand (for more information on load factor, see appendix D). In Figure 7-9, the upper profile shows parking lot lighting, which is off during midday and overnight hours under control of photocells and a timer. The lower profile is for a school at which most power is used in the middle of the day (when its pricing may be higher). Without knowing the time at which each one's peak occurs, they may have the same rate class price even though the cost to the utility to serve the parking lot lights may be significantly lower. Data from interval meters would, however, show *when*

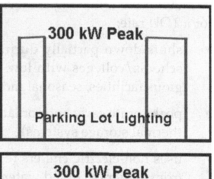

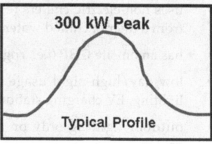

Figure 7-9. Each of these daily load profiles shows a peak of 300 kW, but the parking lot lighting's peak (top) occurs at night while the typical building profile's (bottom) peak occurs mid-day. With older mechanical meters, both accounts would be charged as though their peaks occurred mid-day. *Credit: author*

the peaks occurred for each load and could help a customer find lower power pricing under a TOU rate or competitive power procurement.

WHO MAY BENEFIT FROM TOU?

TOU tariffs are designed to be "revenue neutral," meaning that a utility would not see a change in its revenue if *all* customers in a rate class were switched to TOU pricing. Individual customers in that class may, however, see higher or lower bills depending on when they use the most electricity. Customers having one or more of the following characteristics may be good candidates for a TOU rate:

- shuts down partially during summer (e.g., winter resorts, schools/colleges with low summer attendance, some religious facilities, seasonal industrials).

- peaks quite early and/or late in the day (e.g., electric transit, thermal storage systems).

- uses non-electric chillers (e.g., gas or steam drive), or buys from a district chilled water system.

- has an on-site CHP (i.e., cogeneration) plant.

- low-day/high-night usage (e.g., live theaters, athletic field lighting, EV charging stations).

- outdoor lighting only on at night (parking lots, streets, bridges).

- large data centers that run 24/7; while their on-peak usage will be expensive, many more hours occur during the lower-priced off-peak period, yielding net savings.

- facilities with large PV systems (providing lots of solar kWh during summer peak price periods, instead of buying it from the utility).

If on-site power storage via batteries becomes economical, facilities having it may also profit under a TOU rate.

GET IT 'WHOLESALE' VIA RTP

Unlike TOU, real-time pricing (RTP) is an hourly price based on the utility's (or power supplier's) short-term cost for power. As with TOU, RTP may require use of interval metering. If offered by a utility, it may reflect a mix of its own incremental generating costs (higher during peak times due to use of less efficient generation) and that of hourly power it buys from the wholesale grid. Ideally, RTP should reflect only supply costs, but some utilities mix in a portion of their delivery costs under the theory that loads occurring at certain peak times require greater expenditure for rarely used transmission and distribution capacity.

On average, RTP may offer the lowest annual electricity supply cost—possibly 10% to 20% less than standard rates—but it is also the most *volatile* pricing option. Under RTP, the utility (or power supplier) passes wholesale supply pricing through to a customer (instead of its standard rate and fuel adjustment charge) plus a batch of small adders that cover its capacity, line loss, grid charges, overhead, and profit. The typically lower price stems from the lack of hedging (i.e., financial risk mitigation costs) added by a utility or supplier offering regulated or fixed supply pricing.

Hourly pricing does *NOT*, however, mean hourly billing. Instead, a monthly bill may be based on one of the following:

• a standard load profile for a customer's rate class. An account's monthly kWh consumption is assumed to be distributed across time the same as all members of that rate class (e.g., X% of it occurring from 2 to 3 PM, Y% from 3 to 4 PM, etc.). The total cost is shown in the supply part of the monthly bill. Digital metering is not essential for that process. A customer with an unusual load profile may not, however, realize the savings that may be available if such metering existed.

• on the customer's own hourly loads. Each hourly wholesale price is multiplied by each actual hourly kWh consumption, with the total summed on the bill. Digital metering is required. In this scenario, a customer having a load profile that is the inverse of hourly market pricing (e.g., lower use in late afternoon when hourly pricing may be high) may secure the most savings, though those with fairly flat profiles (e.g., data centers or industrials) may also do well.

In organized markets, hourly wholesale supply pricing is publicly web-posted by the grid operator. If offered a day in advance, pricing for each of the next 24 hours (called the "Day Ahead Market" or DAM) may be available, allowing a customer with load shifting ability to plan its next day's operations based on price. If provided an hour ahead (called the real-time or "Hour Ahead Market" or HAM), automated programming of an EMS may be needed to shift a process on that short notice.

Figure 7-10 shows a year of hourly pricing on its vertical axis (in \$/MWhr) and time and date on its horizontal axes. For 90+% of the time, pricing may be low (in the medium gray bottom band), but may spike by factors of 3 to 10 (or higher) for several hours or days, with little or no warning. During sustained weather extremes (e.g., 2013-2014 northeast Polar Vortex winter, 2011 Texas sustained heat wave), customers on RTP got fiscal whiplash as their supply prices gyrated. Some monthly invoices approached those covering an entire prior season, before falling again for the rest of the year.

While I have not examined RTP in all organized markets, those that I have (in the northeast and Texas) appear to show that RTP supply pricing is lower than either fixed or utility standard pricing in roughly 3 of every 4 years. Unusually high pricing in that fourth year, however, has sometimes nullified much of the savings in the other 3 years. There are also no guarantees against having two "fourth years" in a row.

Before implementing RTP, perform a cost analysis using real (not simulated) interval data and real hourly pricing (available from the utility or grid operator), both for the last year. When com-

2003 Hourly LMP Zone J

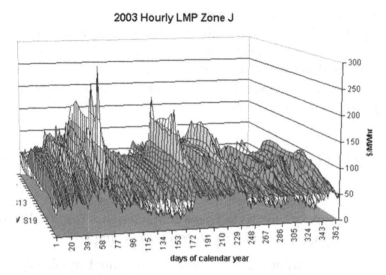

Figure 7-10. 3D chart shows how wholesale hourly power pricing may spike across a year's time. *Credit: author*

pared to actual utility invoicing for the *supply* part of a prior year's bills, an estimate of savings (or loss) will result. Note, however, that there is no guarantee that last year's grid pricing (or a facility's hourly usage) will exactly duplicate itself in the coming year.

In the power procurement industry, it is axiomatic that risk and price run opposite to each other: the more price risk that is taken, the lower the annual cost, in most years. Shifting to RTP must be preceded by consultation with upper management and staff to make them fully aware of the potential risks, rewards, and the price and billing volatility that may occur. Accounting, purchasing, and facilities management departments need to concur on the rate switch. The firm's attorney should review the contract or tariff to ensure that a RTP commitment has an acceptable escape or termination clause.

In deregulated states, some large customers have opted for a mix of RTP and fixed pricing, based on specific months or seasons. Called a "floating price" contract, the fixed price portion of the deal may apply for those months in which historical wholesale pricing has been volatile, while RTP would apply in months whose pricing has been generally dependable. Such a mix may minimize RTP's potential downside while remaining open to its potential upside (e.g., wonderfully low pricing in some of those dependable months). As described in chapters 12 and 13, other types of contracts (e.g., block and index) offer additional ways to mix fixed and RTP pricing to match a customer's risk-reward profile.

WHO MAY BENEFIT FROM RTP?

Many of the same facilities that could save with TOU may also be appropriate for RTP. These characteristics (especially the last two) may qualify a facility to consider trying RTP:

- good DR and/or real-time load management capability.

- large industrial load(s) having flexible scheduling (e.g., kilns, drying, flexible batch production).

- significant (at least 50% of peak kW) low-emission generation.

- flat 24/7 loads (e.g., data center) using lots of kWh in non-peak hours.

- financial ability to accept an occasional "hit" from brief high wholesale pricing.

- senior management willing to accept some risk to achieve lower utility costs.

Here is a bit of wisdom (based on experience) that is worth many times what you paid for this book: during a deep recession, many other types of customers may also benefit from RTP. At such times, grid-wide generating capacity may greatly exceed demand (reduced by an economic downturn), resulting in few (or no) price spikes, even during weather extremes. Switching to RTP at the *beginning* of a recession may then yield several years of lower supply pricing. When the economy picks up, however, a switch back to a lower risk option would be appropriate.

Chapter 8

Equipment/Technical/ Operational Options

REDUCE OR ELIMINATE A REACTIVE POWER CHARGE

Many utilities presently bill customers a "reactive power" charge when their load exhibits low power factor (PF). The charge may be only a few percent (or less) of a total electric bill, with most of it caused by industrial and process loads that may reduce PF. As more smart meters are installed, many more utilities have begun charging even non-industrial customers for PF.

First, determine if—and how much—you pay for PF. It may be charged as a percent below a defined level, e.g., 95%, or as kVAR (kilovolt-amps reactive). To nail down what loads are causing low PF, measure and chart kVAR in 15-minute intervals. Such data may (with utility permission) be available from a utility's meter, or may be measured using a data logger or power analyzer (e.g., Fluke and others make inexpensive units) equipped to do so. Two common sources of low PF are early (pre-1990s) variable speed drives (VSD) and some types of older elevators.

Once PF is known by end use, determine how much correction is needed to cost-effectively raise it. While adding PF correction capacitors (see Figure 8-1) is the usual method, some efficiency upgrades

Figure 8-1. Power factor correction capacitors may help reduce utility reactive power charges.
Photo credit: efxkits.us

that cut kW and kWh may also raise a facility's PF. Facilities where lighting is a significant part of the total load (e.g., galleries, retail stores) may find that upgrading it with high PF ballasts and lamps raises PF sufficiently to zero out (or significantly reduce) a reactive power charge.

In the future, ensure that specifications for new electrical equipment, e.g., large motors (or devices that use them such as chillers), variable speed drives, and lighting, call for high power factor choices, even if that means spending a few extra bucks. Don't allow "value engineering" (translation: buy the cheapest junk and let the next guy deal with the results) to cause higher future electric bills.

CONVERT ELECTRIC LOADS TO NATURAL GAS

Some electric loads may instead be served by natural gas-driven systems. Even where a utility or state blocks cogeneration, it cannot stop conversion of electric loads to gas drives. When some electric utilities tried to do so, they were defeated at the PUC level.

As with CHP, commission a study that shows how much electric load the utility may lose if natural gas replaced electricity to serve large loads (e.g., air conditioning, industrial drying, gas compression). To capture a utility's attention, a large (e.g., a megawatt) potential loss of load should be at issue which could impact the utility's economics (e.g., via a loss of distribution revenue).

Show economic viability (e.g., via an ESCo contract that includes possible financing of conversions to gas) and seek a rate cut to *defer* conversion for a long period, such as 10 years.

Note that some utilities actually reward such demand "destruction" via rebates. Where a utility supplies both power and natural gas, it may actually welcome such a transfer of load from its electric division to its gas division. Consider such issues before seeking such a deferral.

Figure 8-2. A gas-fired air compressor may shift a large electric load to natural gas. *Photo credit: Quincy Compressor Co.*

Question Authority

A few utilities that supply both power and gas have tried to intimidate customers by claiming that "it's not legal to use one utility service to cut the revenue of another." In such cases, "legal" is a misnomer: at worst, doing so could be a violation of a tariff, which is a regulation, not a criminal statute. When I challenged several similar claims, all turned out to be unsupported utility folklore.

In the 1980s, I was confronted by an account rep making that specific assertion. I responded by telling him that he would look like a fool if I challenged his claim before the PUC because the utility has been allowing that action for decades. He said, "Oh yeah? Show me!" I pointed to the ductwork over his head and to its reheat, heating, and preheat coils. The first used electricity, the second used hot water from a natural gas boiler, and the last used steam provided by the same utility.

"When heat is required, the gas-fired heating coil comes on. That air may then be further warmed by a zonal electric reheat coil. Depending on how they are sequenced, one could be cutting the revenue of another. In front of both is a preheat coil that runs on your utility steam. It too may be supplementing or competing with the energy sources feeding the other coils. Are you going to tell me that all such HVAC systems are somehow barred by your tariff? If so, show me that provision."

The account rep grumbled a bit and then conceded the point. To the best of my knowledge, that claim was never again made by the utility. In later years, it offered a rebate program that paid customers for converting their electric loads to instead use the utility's natural gas or steam services.

GET PAID TO CUT LOAD BY
DEMAND RESPONSE PROGRAMS

In many (but not all) states, a reduction in demand upon request of a utility or grid operator is called demand response (DR). When hourly wholesale prices start to spike, they may ask customers to cut load and/or run their emergency or backup generators. Doing so may cut the hourly wholesale price for all customers at that time. Customers may also offer load reductions in advance to compete with output from the power plants feeding into the grid. Some DR programs also seek to limit peak demand in areas where distribution systems are stressed during high load periods.

WHAT DR IS ALL ABOUT

At times of peak grid demand, hourly pricing may skyrocket. In the NY ISO grid in 2000, for example, grid-wide load (see Figure 8-3) generally remained below 24,000 MW about 95% of the time, yielding a wholesale energy-only price at or below $52/MWh ($.052/kWh). But on a hot day in July, it exceeded 30,000 MW. The wholesale power supply price for all buyers in that hour rose exponentially, yielding a price of \$420/MWh (\$.42/kWh). In effect, increasing total demand by that last 25% increased system-wide hourly pricing by about 700%. Such pricing may be passed along to retail customers through a utility's fuel adjustment charge, RTP rate, or via a supplier's floating-priced power contract.

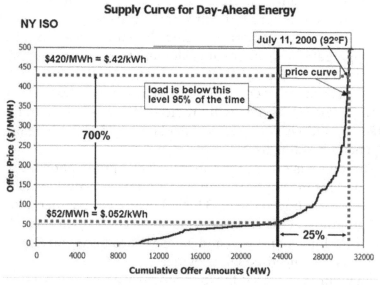

Figure 8-3. Wholesale hourly power pricing may rise exponentially as grid load approaches grid capacity. The last 25% of load in this case led to a 700% increase in hourly pricing. *Credit: NY ISO*

When power customers react to such pricing (or an incentive based on it), that's called "demand response" (DR). When enough do so, the peak load on the grid is reduced, and the price for *all* kWh may drop significantly during those hours. All customers (except those on fixed price supply contracts and those who are not charged a FAC) may then benefit from the lowered pricing.

DR payments range widely, both geographically and seasonally. Some programs are layered so that a customer may participate in several at the same time. As a result, payments (in 2016) ranged from \$.25 to over \$1.50 per avoided kWh and from \$40 to over \$100/kW-yr for pledged capacity (1 kW-year = .365 MW-day). A pledge to reduce demand by 100 kW, for example, may elicit an annual payment of $100/kW-yr x 100 kW = $10,000/yr, even if no reduction request was issued. In nearly all cases, calls for reductions occur during summer, but a request to cut load could be called whenever allowed by a DR program's rules. A few, for example, occurred on the first hot day in April.

Most programs require a participant to provide at least 100 kW of load reduction, but third-party companies (called "demand response providers" or "curtailment service providers") may aggregate loads of small customers to reach that threshold (if allowed by a DR program).

To participate in DR, a utility (not a customer-owned) smart meter must exist. Its data will provide kW loads before and after a DR call, and will be used to calculate avoided kWh against a customer's baseline usage.

To some customers, the paperwork and preparation for some DR programs may appear daunting. Many DR providers exist to handle it for a portion (e.g., 30-50%) of the DR payment. While most programs allow customers to handle DR on their own, those in the PJM Interconnection and New Mexico (as of 2017) instead require use of a licensed DR provider.

In Figure 8-4, we see the load profile of a refrigerated storage warehouse participating in several DR programs. During a DR call, it cut its demand by ~800 kW for several hours. Because the facility was enrolled in a program that paid for committed demand reduction, that action earned it a ~$64,000 incentive. It was also paid for the kWh it avoided during those hours.

Since a DR call may signal when an ISO's grid-wide summer demand is likely to peak, cutting load at that time may also reduce a facility's capacity tag in the following year, i.e., the peak kW upon which a ratchet charge is based. That turned out to be the case for the warehouse, so the facility saved an additional $72,000 in the following year. Note, however, that its *monthly* peak demand did not also drop. This is a good example of why it is essential to understand the components in your tariff and the rules of any incentive programs in which you participate.

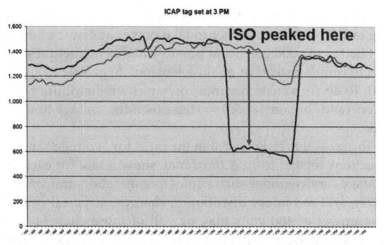

Figure 8-4. In this pair of 24-hour load profiles for an industrial site, the light line is a typical daily profile, while the dark line is a profile during a demand response (DR) call. The facility cut its load significantly, securing DR benefits. The call was coincidentally on the day the grid peaked, so the response also cut the facility's installed capacity (ICAP) kW, significantly cutting its annual ratchet charge. *Credit: author*

WHO OFFERS DR PROGRAMS/TARIFFS?

Many utilities offer demand response or load management options, but rules and benefits vary widely, and from year to year. For a list of those by state, go to http://energy.gov/eere/femp/energy-incentive-programs and click on your state's name. Scroll down to the bottom of that page to find DR and load management programs. That site is updated semi-annually so, before taking any action, verify its pricings and rules with your utility account rep to ensure they are current. Your utility and/or ISO web site should feature details of their programs.

CUT A DISTRIBUTION RATE BY
UPGRADING VOLTAGE SERVICE

Tariff *distribution* rates (both $/kWh and $/kW) may be based on voltage delivered to a meter: the higher the delivered voltage, the lower the rate (if it is shown separately in a tariff). Some tariffs allow a customer to take power at distribution voltages (4,600 to 13,800 volts, and possibly higher) sometimes called "high tension" instead of at "low tension" (e.g., 480, 277, or 120 volts). To do so, a customer must own its own distribution transformers (and/or substation) to step down the voltage to useable levels.

To assess this option, search the tariff for "voltage" and read the sections related to it. A tariff may show a rate for each level of voltage service under such names as generation, transmission, primary, and secondary distribution charges. A typical customer taking power at 480 volts may pay all of those charges, which may be shown separately on the bill, or bundled into one distribution charge.

Where those charges are seen separately, a customer may have the option to request a change of electric service to a higher voltage. For example, an industrial customer receiving 480-volt service may be able to buy from the utility the distribution transformer that serves only its site, or install its own distribution

Figure 8-5. A distribution transformer steps down high-voltage to a level that may be used in a facility. *Photo credit: author*

transformer (which is best done during initial construction, rather than as a retrofit). Doing so may eliminate a secondary demand charge previously paid to reduce voltage from 13,800 to 480 volts. The customer would still pay the primary demand charge (to step down from 120,000 volts to 13,800 volts) and the generation and transmission demand charges.

Cost for new distribution transformers vary but may be ~$25-$60/kVA based on size, type, etc. The payback period may be long (e.g., 10 years), but the cost of the transformer may be deductible as business equipment. Confer with an accountant prior to taking any action. Note that a lower cost may be possible if the utility is willing to sell its existing unit to the customer. That option may be viable when the transformer is located on the customer's property, and does not serve any other customers.

To finance the cost, a utility may offer an on-bill payment option (e.g., a lease) configured so that the customer always sees a savings. Some third-party power providers may engineer and finance such an upgrade through a fixed adder to their supply contract price. In both cases, no major upfront cost would be incurred, and savings would start immediately, all via a paper transaction and changes to utility metering.

Note that some power is lost during voltage reduction. Older transformers may lose 1-2% while modern Energy Star-certified units may lose only .5% (or lower). Once the utility meter is moved to the primary side of a customer-owned transformer, those losses will be borne by the customer. While transformers may last 40+ years, routine maintenance (e.g., annual fluid checks) and insurance liability also shift from the utility to the customer.

One of the best ways to apply this option is during planning for new facilities or major loads. When adding a wing onto a hospital, for example, consider placing it on a separate account and (as part of its construction) including its own distribution transformer. Request a tariff rate that charges only for primary distribution. Reap the savings as soon as the new facility's power is turned on.

PURSUE TRANSMISSION BYPASS

A large facility (e.g., prison, hospital, state university, industrial park) surrounded by open property may have a transmission line crossing its boundaries. As with distribution charges, transmission charges may (if allowed in the tariff) be avoided or reduced by installing a customer-owned substation on its property, and connecting it to the high-voltage transmission system. The facility would remain a utility customer.

A more aggressive approach may be taken by applying section 201 (and others) in the Federal Power Act, wherein a narrow option may exist that allows a customer to connect its own substation to the transmission line on its property that bypasses the utility to directly access wholesale grid pricing. For more details, Google on '"federal power act" + bypass and have your firm's attorney read https://www.ferc.gov/legal/maj-ord-reg/land-docs/rm95-8-0ad.txt and other relevant FERC decisions available from that search term. Exercising such a bypass right may, however, entail a potentially expensive FERC proceeding where the utility is likely to oppose you.

Figure 8-6. High-voltage transmission lines crossing property of a prison. *Photo credit: author*

The goal in this case is not to actually win the ability to bypass, but to instead secure a rate cut as a *quid pro quo* to avoid a conflict and/or setting a precedent that others may seek to copy.

Start by reviewing the existing arrangement with the utility or transmission provider (which may be unrelated to the utility) that allows it to use your property as a transmission path. It may be leasing the rights to do so, have an easement, or have secured it under a utility's right of eminent domain (which allows taking of property rights for the public good). If not a permanent arrangement, note when its term ends and press for a substation (or equivalent discount) as part of renewal of that agreement. Keep in mind, however, that the best time to secure the maximum value from allowing access to or passage through your property is during that initial negotiation, not after a deal is already in place.

If an opportunity to challenge the status quo exists, commission an electric engineering study to assess the economic viability of bypass. Such a study should include attention to any local code requirements regarding on-site facility staff's electric licensing requirements. Some jurisdictions require that a licensed high-voltage electrician be available where high-voltage power is in use. That may entail extra cost or training for existing personnel, or a service contract with an electric service maintenance firm. If the overall numbers make sense, show it to your utility account rep to start a negotiation. Such an analysis may cost tens of thousands of dollars, so some financial risk is involved if the process is not successful. A utility is unlikely to be responsive, however, unless a customer shows that it has the data to make a strong technical presentation in a FERC proceeding.

BUY CHEAPER POWER GENERATED ON-SITE

Many states (including those that have not deregulated retail power supply) allow a customer to buy energy from an on-site non-utility supplier. Ask your utility account rep and check the tariff for specifics.

Options include solar photovoltaic (PV) panels and Combined Heat and Power (CHP) plants, also called cogeneration. Under a long-term (e.g., 15-20 years) power purchasing agreement (PPA) with a developer (e.g., an ESCo), a customer agrees to buy all the energy (electricity and, if produced, heat) from the system. In deregulated states, a PPA may allow the developer to broker the remainder of a customer's electricity supply from others instead of the local utility.

The developer finances and installs the system at little or no cost to the customer. Pricing for the electricity (and heat, if available) is often fixed at a discount off the price(s) that the customer would have paid without the system.

Peak kW output of a CHP system designed to maximize energy efficiency may be only a fraction (e.g., 25-50%) of a facility's peak demand. It may, however, supply 40-70% of total annual kWh if the system runs continuously at its design output. Depending on size, facility operations, and other factors, a PV system may also provide most or all of a facility's annual kWh, but may not significantly reduce a facility's peak demand, unless the facility's peak occurs near noon.

Some customers have had second thoughts after pursuing such installations. If, for example, A CHP system fails at the time a facility is hitting its peak demand, the developer should cover any resulting demand and ratchet charges. Many PPAs, however, do not address such issues.

Before signing it, have the contract reviewed by an energy consultant experienced with on-site generation and PPAs. Scrutinize its pricing formulae for purchased power and heat to ensure acceptability. Meter the purchased power in 15-minute intervals, and separate the payment for energy (kWh) from that of the capacity (kW) provided at the time the facility peaks. Unless a PPA contains a proviso to make the customer whole if the system does not fully perform during a facility's peak demand, do not buy electricity at a fixed rate near a facility's average electric rate. Doing so assumes the system will always lower peak demand (as seen at the utility meter)—by an amount equal to its capacity—at the same time that the facility's kW peak normally occurs.

NEGOTIATING COGENERATION

Where cogeneration is not already allowed under an existing rate option, negotiation with a utility may be necessary. A good source of guidance is John Studebaker's book *Electricity Purchasing Handbook*. It offers a soup-to-nuts discussion for persuading a utility to allow a customer (or its developer) to supply on-site power.

Before initiating negotiations, determine if other CHP plants are operating in the utility's territory. Review the database of CHP plants at https://doe.icfwebservices.com/chpdb/ and consult with your state PUC's consumer advocate to check for helpful precedents.

Issues related to on-site power systems are covered in more detail in Appendix G.

CASH IN ON OPTIONS FOR
USING ALTERNATIVE ENERGY SOURCES

Some states offer financial incentives to produce on-site power from low- or no-carbon sources such as solar photovoltaic (PV) panels, fuel cells, or CHP. Rate options vary across states and among utilities within states. Find details on many incentives at www.dsireusa.org.

Net metering (NM) is the most common opportunity. NM tariffs allow a customer (or its on-site developer) to use the local power grid like a battery, pushing power generated (but not used) at a facility into the local distribution grid, and taking it back at other times. Many residential PV systems create more kWh than needed at a home in a month, and the extra is credited for use in later months. Ideally, a customer then pays only his fixed charges for electricity. When applied for a C&I customer that pays a demand charge, however, only the energy component of the bill may be affected. As of 2017, several states have begun scaling back their support for NM, so it's best to assess continuity of such programs before investing resources in their pursuit.

Care is also needed when projecting savings and determining their payback periods. The economics of alternative energy generated at C&I facilities may require an hourly load analysis. Depending on the tariff, the value of excess on-site power may be reimbursed at the utility's own power supply rate, or the combined supply and delivery energy rate, or at a wholesale (not retail) solar power price, none of which may include demand. Where demand is a significant part of a total electric bill, net dollar savings may be lower than expected unless a proposed system is designed to significantly cut peak demand (e.g., via specific PV panel orientation to do so). Such a design may, however, then not produce as many kWh as a typical PV system. Only when a PV system is equipped with motors and solar trackers that tilt the panels to follow the sun may this problem be minimized. Such systems may, however, cost much more (e.g., 40% to 70%) to install and maintain than fixed panel systems. If large scale battery power storage ever becomes cost-effective, it too may be employed to address this problem, sometimes called the "solar dilemma."

Figure 8-7. Roof-mounted photovoltaic (PV) panels. *Photo credit: Village of Croton on Hudson*

Note also that tariffs or PUC regulations may impose limits on such systems. In some states, a PV system over 1 or 2 MW may not be eligible for net metering. A utility may also limit the total renewable kW it will allow on its system (or some parts of it). On-site generation may trigger a rate class change, and (for CHP or fuel cells) require that certain minimum thermal efficiencies be maintained to avoid being switched to a less favorable rate. Some states are (in 2017) scaling back or replacing their net metering programs with less lucrative—and/or more complex—rate structures.

On the plus side, renewable power may allow a system owner to sell Renewable Energy Certificates (RECs) that represent the environmental attributes (e.g., zero carbon production) of that energy. A robust wholesale market exists for RECs, which are traded every day like other commodities. Most states have Renewable Portfolio Standards (RPS) that require utilities to either generate a portion of their power from renewable sources, or else buy RECs from other power suppliers that do so. The utility then owns the environmental attributes for that power, meaning that a customer generating it cannot count it against its own carbon footprint.

A few states have similar programs that include low emission power production (e.g., from fuel cells) to be reimbursed through Alternative Energy Credits (AECs). Note that pricing for RECs varies widely by state, and may be much higher for power produced from solar PV than from other renewable sources (e.g., wind).

In addition to (or in lieu of) RECs, some utilities offer Feed-in Tariffs (FIT) that pay for renewable power at rates several times higher than its on-site value. If, for example, a customer's average power price is $.10/kWh, a utility may pay it $.30/kWh for the PV power it produces. The utility also secures the environmental attributes for that power. FITs have been common in Europe and other nations, but are presently offered by only a few U.S. utilities.

Section III: Changing The Rules

Chapter 9

Negotiating with Utilities

Several of the options we have discussed may require working with your utility. Some facility managers are reluctant to do so out of fear their lights will suddenly go off, or from past frustrations when dealing with their utilities. Recall that even though a utility has a local monopoly on power delivery (and maybe on supply) it is, in essence, just another vendor. Following are tips that may make the task easier.

Recall that four different types of utilities operate in the U.S.:

- Investor-owned (IOU)
- Co-operative (co-op)
- Municipal (muni)
- Power Authority (authorities)

When it comes to negotiating an opportunity, each type has its own agenda, so it pays to understand what may (or may not) work for them. At all times, flexibility and patience are essential to success.

To get something it desires, a customer must offer something of value to the utility. Here are a few options to which a utility (which is just another form of business) may respond.

- Offer a way to cut its costs, e.g., take over maintenance of utility equipment on your property.
- Demonstrate a viable option for cutting its revenue, e.g., installing on-site cogeneration.
- Raise its revenue, e.g., by adding new load that could otherwise be located outside its territory.

- Lower the utility's peak load; e.g., through installation of a thermal storage system.

- Maintain its revenue, e.g., by not moving your business out of its territory.

- Increase load flexibility, e.g., by making some load interruptible upon request.

- Help it get something it wants in a rate proceeding, e.g., by changing your position or opposition.

- Enhance its stature before a PUC; e.g., propose a way for it to look good without spending (or giving up) money, such as making a customer's interval data readily available to it at no charge.

Success in such efforts may be enhanced by working through (or forming) an energy users' group composed of other C&I customers and appropriate trade allies (e.g., ESCOs). See appendix I for a discussion of that option.

IOUS: OFTEN THE TOUGHEST NUTS TO CRACK

An IOU may have a narrow mindset: "I already negotiated with the PUC. Anything I give you raises the rates of others." This is not necessarily true. If a customer's proposal looks promising, an IOU may instead find ways to reduce or shift its costs, or adjust shareholder dividends, or accept a ROE slightly lower than allowed by a PUC (usually between 9% and 12% on invested capital). Unless a utility is quite small compared to the proposed change, the actual impact on it may be less than a rounding-off error in its annual budget.

To make the case before a PUC for a tariff or other change, An IOU and/or a customer need to work together. Some PUCs have dollar limits below which a utility and customer can come to an arrangement (e.g., to purchase utility equipment, such as a transformer) without PUC involvement. Any action that could shift costs among ratepayers (e.g., by giving a large customer a special rate), however, is likely to require PUC approval, though such action need not always be public.

In all cases, a customer or group must come well prepared to a negotiation, with numbers that back up its requests. Find allies (e.g., ESCOs, trade organizations) that support your position and bring their reps to the table with you. More details on tactics follow below.

Some large customers or customer groups have tried to gain leverage over their IOU by proposing that their local government take over a portion of its delivery system (i.e., its T&D assets) through "municipalization." This process seeks to force a sale (e.g., via eminent domain) of utility assets to a town or county and convert them into a muni. Having closely watched several such efforts, I do not recommend pursuing it. Unless a utility has exhibited seriously incompetent service (e.g., chronically failing to expeditiously restore service after outages), sufficient political will may not exist to take on electric utility service as a government function. In every case I have observed, the main winners were the attorneys that collected large fees regardless of the outcome.

While some leverage may develop during that process, the usual result is instead side deals with the main contenders that could have been achieved quicker and at lower cost through standard negotiations. All other options need to be fully exhausted before engaging in what may be a long expensive fight on turf IOUs are accustomed to treading (e.g., lobbying, public relations, etc.).

MUNIS AND CO-OPS MAY BE EASIER

Publicly owned utilities (which have no shareholders or profit margin to maintain) may be more community-oriented, and open to helping local businesses and institutions. Customer proposals that add or maintain jobs, attract new power customers or load, help stabilize rates, and/or expand the tax base may be welcomed.

AUTHORITIES ARE A MIX

While also non-profit, many are quasi-state agencies that act slowly and may be unaccustomed to dealing with retail customers. Approaching one of them as part of a group (creating an appearance of potential political clout) may be a useful way to get attention.

When approaching a utility—regardless of type—a well-defined proposal is essential. Entering a meeting with only vague demands or unspecific ideas will simply reduce your credibility (both present and future).

POSSIBLE MECHANISMS FOR A RESOLUTION

To accommodate a customer, some utilities have entertained one or more of these options.

- A common way to give a customer a rate cut has been to create a temporary "pilot" or "experimental" rate specific to its (or a group's) characteristics. It is then routinely renewed, unless protested by others. One version provided a risk-free RTP or TOU rate wherein total annual cost will be no higher than would occur on standard rates (any overage is rebated), with any savings during the tariff's term being kept by the customer.

- To keep industrial jobs in their communities, some publicly owned utilities have changed the boundaries of their territories so a customer was able to access a lower rate at an adjacent utility. That adjacent utility must agree, and some cost to alter distribution lines may be involved.

- Some have allowed buy-through to the wholesale grid, in essence creating de facto deregulation for specific customers. The customer may then pay a form of wholesale RTP for its supply.

- Through creation of a "green" tariff, a utility allows renewable power bought by a customer from a third party to be shipped (i.e., "wheeled") through its system to the customer. This is similar to retail deregulation, but limited to very large customers, only a few suppliers, and only for renewable power.

- To give some customer access to economic development and/or business incentive rates, utilities (with PUC approval) have loosened definitions of "customer," "account," site," "property," and other terms that otherwise limited such access.

- To attract or retain desired types of customers (e.g., data center, medical research) that flatten utility load and/or bring well-paid jobs to the community, incentive rates have been offered that run for a decade or more.

Such options may be narrowly defined "special provisions" or "riders" written in ways that conform only to certain customers without naming them (e.g., "...applicable to all customers with peak demands exceeding 5 MW north of 50th Street that have continuously maintained an account in that area since 1959"). For hints, ideas, and wording, review such provisions and riders already in the tariff. See also options listed in Appendix H.

NEGOTIATION IS NEVER A SIMPLE OR EASY TASK

Negotiation is a process, not a battle. For a deal to work, both sides must come away with something. Having an experienced consultant may go a long way in such discussions. Using in-house personnel or a company attorney may be cheaper, but doing so could end up costing more than doing nothing if a serious error is made. Don't think that someone lacking experience in energy regulatory or procurement activities can "learn while doing."

Unless already well versed in utility law and tariffs, an attorney should work with a consultant to help with strategy and negotiations. To ensure an expeditious result, seek a reduced hourly rate (e.g., 60-75% of the consultant's usual rate) but offer a percent of secured savings (e.g., 10% to 25% for 2-3 years) so the consultant has some "skin in the game."

Several books may be helpful when preparing a proposal:

- *Getting To Yes* by Roger Fisher and William Ury

- *Guide to Purchasing Electricity and Gas* by Paul Cunningham and David Burrell

- *Utility Negotiating Strategies For End-Users* by John Studebaker.

The key to starting a discussion is to offer a valid benefit in exchange for something you desire: don't expect to succeed merely by asking or demanding. Once a proposed deal is developed, rehearse its presentation with a co-worker to test its credibility and clarity *before* meeting with the utility. Sharpen as needed by distilling your opening to a 30-second "elevator" statement (that's about 50 words). You may get only one good first shot.

Keep detailed records of all meetings and calls with their dates and attendee lists, including contact information (phone, email, address) and titles.

Persistence is key: most customers fail simply by giving up. It is a standard utility tactic to simply draw out a discussion through delays, postponed meetings, changing personnel handling the issue, and other seemingly childish ways to avoid a conclusion. Doing so may run up such high legal fees that the customer chooses to cut his losses and drop the effort.

While patience is a virtue, however, don't hesitate to file a formal complaint with the PUC (or other body overseeing the utility) if, after several months, intentional delays continue. Some utilities hate dealing with such agencies.

Remain open to being "bought": utilities have ways to avoid changes and precedents that others may later seek to exploit. See the boxed story for an example.

How to Find $50,000

To get a customer to drop out of a nasty rate proceeding in which its opposition to a rate hike was having effect, a utility offered (to the customer's CFO) to "review your past bills for errors," indicating that it might find up to $50,000 in "mistakes."

To do so, however, it wanted the customer "to work with" the utility in the rate proceeding. The "billing errors" were corrected and $50,000 in "adjustments" appeared on the next electric bill. As an act of good faith and to ensure future good relations with the utility, the customer dropped out of the proceeding.

SWEAT THE DETAILS OF A PROPOSED DEAL

As with any commercial arrangement, the result may contain provisos designed to limit its value. Read proposed wording carefully, with an eye to both the "what ifs" and these potential pitfalls.

- Term: for how long does the deal last? One year is too short to bother, and 15 years may be too long if a significant commitment is involved (e.g., deferring on-site generation options).

- Potential penalties: Does the agreement end or is the customer penalized if, e.g., its annual usage drops below a defined number? Is there a penalty for early termination?

- New or hidden costs or rules: could any aspect of the deal result in new or additional fees or costs such as balancing or line loss?

- Future involvements: are any "political" promises involved, e.g., choosing not to participate in its next rate proceeding?

- Customer name: will it be used in any way, e.g., public announcement of the deal or implied endorsement of the utility?

- Name or ownership change: is the deal lost if the customer is bought or merges with another?

- Financial position: might a sudden large "refund" (that looks to the IRS like undeclared income) negatively impact the customer's taxes?

- Operating restrictions: does the deal block use of existing or future equipment (e.g., on-site power) or require involvement in utility-controlled programs (e.g., demand response)?

- Site limitations: is the benefit limited to only one customer site in a utility's territory? Is the deal terminated if the customer relocates within that territory?

If all else fails, consider a formal complaint against the utility. Most PUCs have simple procedures for filing them, perhaps through an on-line form. Utilities are required to respond within defined time limits. In some cases, that is sufficient to start or accelerate a stalled dialogue. If not, it may be escalated to a point where the parties make their cases before a PUC mediator or (in rare cases) before the PUC's members.

Chapter 10

Intervening to
Help Write the Rates

INTRODUCTION

Utilities are regulated by appointed or elected PUCs or boards. They create or alter rates during formal proceedings managed by PUC staff and overseen by an administrative law judge (ALJ). Typical participants include the utility, government agencies, lobbyists, ESCOs, customers (and groups of them), environmentalists, and others.

To become a participant is to "intervene" in a proceeding. Intervening is *NOT* lobbying: many non-profits do it. It requires achieving "standing," i.e., making a showing that you represent a party that could be impacted by the proceeding. A request for standing is usually made to, and approved (or denied) by, the ALJ. Doing so may involve little beyond a letter showing that you or your group has electrical accounts worth $XXX,000 dollars a year.

Another intervenor may try to dissuade a grant of standing if it sees that entity as a possible opponent. That rarely works, but may require a response by an attorney. Participation may, in any case, involve engaging an attorney experienced in regulatory proceedings to represent a customer (or group), but customers may also participate directly.

Before becoming an intervenor, review past PUC proceedings to see what groups are already active in such efforts. Your firm may choose to instead join one and gain the benefit of its experience and position.

Two types of proceedings are common:

- A customer (or group) requests a limited proceeding focusing on a specific desired change. It may be opposed or supported by others such as vendors, suppliers, and special interest groups (e.g., environmental, housing, union).

- A general rate proceeding requested by the utility that may impact all ratepayers. The goal of customers may be to limit an increase or reduce the rate for their class (and/or seek special treatment for their type of facility or industry).

UNDERSTANDING RATE PROCEEDINGS

When asked if intervening makes sense for large customers, a utility CEO once said: "If you don't sit at the table, you won't be fed." It makes no sense to complain about a rate hike *after* it happens: the time to be heard is while the deal is being made.

Keeping current on upcoming proceedings is not difficult. Many PUCs post notices and status reports at their web sites. The best way to monitor a proceeding is to become an intervenor and keep track of the emailed correspondence and documents circulated to all participants. If too burdensome, designate an employee or find a consultant to do it for you.

In a general proceeding, the utility may present its "revenue requirement," i.e., how much money it needs to continue reliably providing power. PUC staff and participating parties look for ways to cut it. The utility may also offer a "cost of service" (COS) study of how it allocated its proposed rate hike among classes based on load profiles, growth rates, special needs, etc.

During such talks, the parties present arguments, studies, and pleas on how to reduce the request or re-distribute costs across the rate classes. Discussions are confidential during this process, though leaks (especially by politicians seeking to influence the outcome) are not unusual.

Economic fairness based on costs of service is a general goal, but ratemaking may shift costs to fulfill a variety of other demands. Most residential customers are (for example) also voters, and industrials may get breaks to maintain jobs. All may end up paying for programs desired by the PUC (or those that appointed its members) such as energy efficiency, economic development, low income assistance, and/or renewable energy.

In your first rate proceeding, seek or propose options that are not an uphill battle. Consider what small specific actions could benefit you and/or your allies. Gain experience (and credibility) before delving into contentious or protracted efforts less likely to succeed.

To get you thinking, review the wish list in appendix H. Many items cost little (or, in some cases, almost nothing) for a utility to provide. A few of them might even cut a utility's costs and/or help elicit utility participation on your behalf against PUC staff or other intervenors.

In all proceedings, consensus among all parties is desired. However, a deal (called a "settlement") signed by the majority or just the major parties often wins the day. Once approved by the ALJ, it becomes public (but not before). Those not satisfied may pursue litigation based on a claim of improper procedure or bias, but they rarely succeed.

Public hearings on the proposed deal may be held, but most are merely for show and venting of complaints. Very few result in changes to an existing settlement developed behind closed doors. An ALJ or the PUC's board may make changes to a proposed settlement, but few major alterations occur unless it is seriously flawed.

Persistence in this process is critical. A lot may happen in the last few weeks of a year-long proceeding. A critic once likened utility ratemaking to a bar fight wherein "the winner is the last drunk left standing."

TO SUCCEED, DON'T GO IT ALONE

Interventions are a gamble: a participant may spend tens of thousands of dollars in fees for legal and expert witness support and save nothing. On the other hand, a good result for a large customer (or group) might save millions over several years if it reduces an otherwise large requested rate hike. Here are a few "lessons learned" that may help you avoid shooting yourself in the foot.

- Before doing anything, secure from your upper management clear support and authority to participate in a proceeding. It pays to know (in advance) of any hidden relationship (e.g., memberships on a corporate board) between a utility and the customer's ownership or leadership.

- To avoid an appearance of self-serving, join or form a group of allies in the same class or position. Doing so helps establish standing and share expenses.

- Coordinate a strategy and budget to run for a full year. Proceedings may take that long to bear fruit.

- Use an attorney or law firm with experience in regulatory affairs. Support it with testimony from a rate expert with verified experience in utility negotiations (not someone who has just read news articles or internet downloads, or an academic lacking experience in the real world).

- Review the existing tariff and prior rate settlements for clues and options (e.g., special rates given to attract or hold jobs, or for businesses desired by the community, such as manufacturing).

- To get what you want, don't feel limited to PUC rate proceedings. Sometimes a lawsuit against a regulatory body may provide the necessary leverage. Most states have laws (generally called "administrative procedures acts"), requiring their agencies to follow their own rules and other laws. A failure to do so (e.g., by favoring some parties in a proceeding) may allow you to seek a court order forcing them to comply. The object is not necessarily to win in court, but to get the offender to agree to something in exchange for dropping the suit before it gets to court. A "writ of mandamus" (Google it) may be an option to force compliance with an existing law or process.

- Forget abstract "principles." No one has ever paid a bill with a principle. Success is achieved by cutting a present cost, avoiding a future cost, or winning a refund larger than the cost of participation.

There are no guarantees that intervening will net anything beyond headaches and unrecovered expenses. It's a lot like the lottery: you may not succeed, but you "gotta be in it to win it."

Section IV:
Competitive Retail Power Procurement

Just as there may be opportunities for savings under tariffed pricing, so there may be options for lower costs where IOUs have been deregulated. Figure 2-4 shows the states that allow competitive purchasing of electric *supply* (but not *delivery*). Among the C&I customers in the shaded states, many have been buying their supply from competing suppliers for many years.

To understand how this is possible—and how to take advantage of it—a potential customer needs to understand how it happened, and how that competitive landscape continues to shift and evolve.

Chapter 11

Buying from Competing Electricity Suppliers

INTRODUCTION

When electricity is purchased from a utility, the price covers many services, of which supply (i.e., generation) and delivery (i.e., high-voltage transmission and local distribution) are the largest parts. Prior to the 21st century, a retail customer had to buy both services from the utility under regulated tariffs. Depending on energy market conditions, and what a utility may charge for delivery, 30% to 60% of a total electric bill may be supply.

Most C&I customers (over 80% in some areas) behind deregulated IOUs buy the supply portion of their electricity service from non-utility suppliers. Even where a state has deregulated its IOUs, however, not all customers can access this opportunity. Neither municipal (i.e., city-owned) nor co-operative (i.e., rate-payer-owned) utilities were required to participate in deregulation.

In all cases, supply is the *only* part of the bill that may be affected through retail electricity procurement (except in NY, where sales tax may also be affected). As will be seen below, the competitive procurement process offers many ways to shave supply pricing.

Even when savings may be small, some customers prefer the stability of non-utility power pricing over gyrating utility pricing. Others like the option to buy power from environmentally acceptable sources (e.g., wind) instead of the (perhaps) less desirable sources used by the utility.

Where allowed, dozens of retail power suppliers now active-ly sell power to millions of end-use customers, of all types. An entire cottage industry of brokers, consultants, and aggregators (hereafter collectively called "specialists") has grown up to help customers buy competitively priced electricity.

From a practical standpoint, C&I customers that consume at least $100,000 a year of electricity (total annual electricity spend for one or a few accounts), may get the most value out of this process. Smaller customers may also find value where utilities offer clear benchmark pricing ("standard offers") against which competing prices may be compared, or by buying through an aggregation program with others.

Competitively priced power may be purchased under contract in a wide variety of forms, for customer-specified term lengths. While many buy their supply at fixed $/kWh rates, others do so based on hourly wholesale grid pricing, or combinations of those options. Some have combined power procurement with energy efficiency upgrades, using the savings to finance them.

Buying non-utility supply entails financial risk. In some cases, utility supply rates have dropped below those of fixed price power contracts, leaving some customers with a relative loss instead of a savings. To help them navigate the process, many C&I customers engage specialists to help them do so more professionally.

For a variety of reasons, retail deregulation is unlikely to spread to other states. Where it does exist, the rules, wholesale markets, and pricing structures continue to evolve. A few states, for example, now allow their IOUs to discontinue supplying power, instead limiting their main service to delivery. Customers are then either transferred to a competitive power supplier chosen by the utility, or are required to contract with a supplier on their own.

HISTORICAL BACKGROUND

In 1992, changes to federal law and regulation opened access to power transmission grids (previously dominated by large utilities) so that others—munis, co-ops, IOUs, IPPs, and very large customers—could competitively buy and sell (i.e., "trade") wholesale power among themselves. Doing so fostered creation of markets and merchant generators that could compete at the *wholesale* level. As the concept of electric deregulation became popular, states began to look for ways to bring the same capability down to the *retail* level, allowing end use customers to buy power from non-utility suppliers.

In 1998, California (CA) took the lead in deregulating its 3 IOUs, allowing their customers to buy power competitively. Several other states (e.g., New York) followed shortly thereafter. In 2000, however, some large wholesale power suppliers (among them Enron) took advantage of vulnerabilities built into CA's wholesale power market. Two years of rolling blackouts, huge price spikes, and—eventually—a recall of the governor (at the time, Gray Davis) resulted. The negative fallout from that experience stunted the growth of deregulation in the US: Montana closed its market, and a half dozen others followed by creating barriers or limitations.

By 2010, all states in New England (except Vermont), the mid-Atlantic region, plus NY, TX, IL, OH, MI, DC, and OR allowed all or some IOU customers to shop for their power supply. CA later re-opened a small portion (less than 5%) of its electric load to competitive power procurement.

In 2016, the desire to secure power from off-site renewable power producers slightly advanced that process. NV allowed several large casino operators and data centers to utilize that state's unused retail procurement law to buy their electricity from competitive suppliers. A few IOUs in still-regulated states began (via new "green" tariffs) to allow several non-utility suppliers of renewable power to "wheel" (i.e., transfer) it through the IOUs' systems to customers having contracts with those suppliers. In all cases, however, power *delivery* remains a regulated monopoly.

As you read further, keep these two axioms in mind:

• Whatever happens in wholesale power markets eventually trickles down to the retail market. Accessing competitively

priced power may offer savings, but it may expose customers to the volatility of the wholesale market, which may be
greater than that seen in stock markets.

• Much of retail power procurement is like wholesale procurement in miniature. Many of the same techniques may be
used.

HOW THE PURCHASING PROCESS WORKS

Each state's PUC web site lists non-utility suppliers licensed
to serve in its jurisdiction, along with their contact information.
Customers may contact a few suppliers directly and request pricing, or pursue a request for proposals (RFP) to many suppliers,
or use a specialist to handle their procurement. For a commission
(typically ~1% to 3% of the value of the deal), a specialist will
use competitive bidding procedures to secure the best non-utility
supply price for the customer. A good specialist also reviews the
proposed contract, called a power purchasing agreement (PPA),
for issues that may require negotiation or alteration *before* the customer signs it. Called a "bilateral" contract, the utility has no involvement in the PPA: it is whatever the two parties have agreed
to, regardless of the utility's supply tariff.

After a PPA is signed, the new supplier is responsible for
notifying the utility of the switchover prior to the next reading of
the customer's utility meter. Some utilities require only 3 days,
while a few require notification at least 15 days before the next
meter reading. If not properly handled, the switchover may be
postponed until the next meter read date, which may void the
quoted contract price.

Once the customer starts receiving power under the PPA,
the supplier uses the utility's meter readings to calculate its supply bill. Except in rare cases (e.g., switching to hourly pricing
requiring installation of a digital meter), no changes are needed
to either wiring or metering. The utility remains responsible for
maintaining its distribution system: if there's a power outage or

other problem, customers call the utility, not the supplier.

When a customer buys power under a PPA, he may receive a bill from that supplier and a separate bill from the utility covering delivery, metering, and other charges found in a typical utility bill. Many suppliers have arrangements with utilities wherein their charges are instead "consolidated" as a line item onto the utility delivery bill. The customer pays the utility, which then passes the portion related to supply to the non-utility supplier. The supplier's price may be seen on the utility bill, but the terms and conditions governing it will be found in the PPA, not the utility tariff.

Where deregulated IOUs continue to offer supply, they become the "provider of last resort," buying (and/or generating) wholesale power and re-selling it to customers who prefer to get their supply from their utility. Pricing is then governed by the supply tariff regulated by the PUC. Even where deregulation has been in place for over a decade, up to 20% of C&I customers have continued to buy their supply from the local utility.

Where tariffs are simple (e.g., a fixed $/kWh supply rate), a utility may publish a forecasted $/kWh "price to compare," "price to beat," or "standard offer" to allow easy comparison to a fixed price from a non-utility supplier. Such standard offers may be good for short periods (e.g., 3 months) or up to a full year. PUC web sites may list comparisons of fixed standard offer rates for residential and small commercial customers on a monthly basis. C&I supply tariffs may, however, be sufficiently complex that standard offers don't apply. In such cases, an expert eye may be needed to project potential savings under a PPA.

As more smart meters are installed, IOUs are (with PUC approval) lowering the peak kW demand threshold of those who may access standard offers. Starting in 2006, some IOUs began to require that a customer with an annual peak demand exceeding 1 MW (i.e.,1,000 kW) that wished to continue taking supply from them had to do so based on hourly pricing, instead of the simpler tariff pricing paid by smaller customers. Over the next decade, that threshold has in some areas been gradually reduced to 300 kW (or lower), pushing many more customers to take either hourly pricing from the utility or switch to a non-utility supplier. Other states have been similarly narrowing access to standard offer pricing, gradually limiting it to residential and small commercial customers (i.e., those not metered for peak demand).

Savings under PPAs (relative to utility supply pricing) may—or may not—occur. Pricing from non-utility suppliers tends to follow that seen in regional wholesale power markets. They operate like stock exchanges, with similar (or greater) price volatility. When wholesale prices drop, utility supply pricing may not do so for months (or years), but competitive retail supply pricing may drop quickly (unless fixed in a PPA). However, when a regulated price drops (e.g., due to a reduction in utility fuel costs), a customer with a *fixed-price* PPA may then see a loss relative to it. While many customers have saved 2% to 20% relative to tariff supply pricing, buying competitively priced electricity *does not guarantee savings relative to utility supply pricing unless so stated in its PPA.*

PREPARING FOR RETAIL POWER PROCUREMENT

A customer may handle all aspects of retail power procurement on its own. For customers with a relatively flat organizational chart, or a facility manager empowered to sign contracts without involving others in his firm, the process outlined below may be overkill. For others, all (or a subset) of the following steps may be appropriate, especially the first time competitive power is to be purchased. Others may instead choose to use a specialist to guide them. Doing so may avoid pitfalls while providing training for future competitive purchasing.

Experience with first-time customers has shown the need for preparation prior to seeking pricing. Unbundled power services, competing suppliers, sophisticated power products, and price risk may be foreign concepts to those handling utility billing and procurement. Some discussion and knowledge transfer may be needed to ensure a smooth transition and a trouble-free contract.

Power procurement may be approached as a corporate opportunity, and not just another office task. Unlike buying utility power that offers few options, buying power on your own terms may offer a variety of ways to positively raise a company's (or energy manager's) visibility. Failing to consider how that may be done—*before* embarking on the process—is like leaving money on the table.

Prep time and effort are roughly proportional to the complexity of an organization, its geographic spread, and the number of electric accounts. If only one account exists, planning and bidding may be successfully completed in less than 2 months, so feel free to skip ahead to Chapter 12. When a firm has many departments that must be involved or consulted, or more than a dozen accounts, or the accounts are behind multiple utilities, extra time and planning (perhaps 3 to 6 months) may be needed. Trying to rush faster may result in embarrassment, loss of credibility, and/ or higher electric bills.

Step 1 assesses available options and suppliers in states where a firm has facilities, typically by reviewing utility and PUC web sites. Some facilities may be behind municipal or co-op utilities that are not open to competition. In other cases, the amount of utility load that is allowed to participate may be limited (e.g., Michigan, California).

Retail electricity suppliers licensed in each state may be found at PUC websites under pages with titles such as "Power To Choose," "Retail Electricity Providers," "Licensed Retailers/Suppliers," etc. or by entering similar language in the site's Search field. Note that all suppliers may not serve all IOU territories in a state, or all types of customers. PUC web sites typically feature ways to sort through them to find those appropriate to a company's needs.

Step 2 is an assessment regarding how the firm presently purchases power:

•   Which facilities are served by which deregulated IOUs?

- Are their accounts with the utility or as sub-metered tenants (who cannot buy power on their own)?

- Who handles utility billing at each facility?

- Do any facilities have special relationships (e.g., off-tariff discounts) with their utilities or local governments that could conflict or be lost if power was taken from a non-utility supplier?

- Are any facilities presently purchasing electricity from non-utility suppliers?

- If so, who manages each of the contracts and when are they up for renewal?

If facility or purchasing personnel at dispersed sites are presently allowed to pursue power procurement on their own, immediately issue a memo suspending that ability until the central office has nailed down its purchasing policy and procedures. Otherwise, it may be necessary to wait months (or years) for uncoordinated contracts to run out, or find ways to unwind them (e.g., by paying termination penalties) before proceeding.

Step 3 involves the decision-making process:

- Who will determine the main goals of the power procurement effort, e.g., one executive, or a committee?

- What departments will have input when crafting those goals and choosing a supplier?

- How will the bidder list be developed: word-of-mouth recommendations? Issuing a request for information (RFI) to all suppliers? Accepting recommendations from a specialist?

- What department and person(s) will oversee a company-wide power contract?

- If a consultant is to be retained to assist the effort, how will he/she be paid, and from what budget(s)?

While most facilities are focusing on cutting their electric bills, consideration should be given to the following additional goals:

- Should this process be used to polish the corporate image (or fulfill an existing mission statement or commitment) by buying some or all power from renewable sources (e.g., wind)?

- Does a desire exist to integrate the firm's marketing with its power procurement (described below under "Marketing")?

- Could savings from procurement be used to upgrade energy-using equipment (e.g., lighting, HVAC) that could further cut the firm's overall energy bills?

- Is there a need or desire to fix a long-term price, even if possibly higher than presently offered by the utility, in order to create budget certainty for several years?

- Do any sites plan to install on-site generation, or make any significant changes that could impact electricity usage in the near future (e.g., installing solar PV panels, adding or vacating space)?

- Does the firm have any existing relationships with energy suppliers, e.g., for natural gas, that it may wish to exploit, e.g., by combining with electricity procurement?

To answer those questions and manage the process, some—or all—of the following departments may need to be involved (at least initially) in the procurement process.

- Purchasing

- Facilities Management

- Legal

- Accounting

- Leasing

- Public Relations

- Environmental/Sustainability

- Marketing

Failing to do so may result in unforeseen problems or internal dissension. To coordinate such personnel, form a working group with one person (usually from Purchasing or Facilities) taking the lead. That person chairs the group and incorporates input when developing its RFP, preferably with assistance of a specialist. The group may meet once in person to learn how competitive power procurement works, and then communicate online to review the draft RFP. A subset of the group may then be involved in reviewing bids and choosing a supplier.

Let's look at how each department may be involved.

Purchasing

When buying products for a company, the Purchasing department is usually in charge: it has experienced personnel, contracts, and procedures in place. But few purchasing agents have much experience when it comes to energy, other than liquid fuels (e.g., fuel oil and gasoline). Some may have competitively purchased natural gas, but not electricity.

Facilities Management (FM)

FM should have a grasp of major electric loads and patterns. Its personnel should also be aware of any impending changes that could impact electric usage and demand (e.g., installing solar panels). A significant change in usage, without informing a supplier, could trigger penalties in a power contract.

Legal

The Legal department will review proposed contracts to ensure that all provisions are acceptable and risk is minimized. It should be responsible for checking on any potential conflicts with existing arrangements (e.g., a special utility rate or property lease). Note: if an option exists to choose the attorney or law firm that will review the contract, seek one with prior experience in regulatory affairs and/or energy contract negotiations. Too many good deals have faltered when an inexperienced (and/or ignorant) attorney has tried to use the negotiating process to seek "perfect" wording. In the end, learning to live with as much wording as possible in the supplier's standard contract is a much better course.

Accounting

If not apprised in advance of a change of energy suppliers, accounting personnel may be confused when they start receiving electric bills from a non-utility supplier. This is especially important where multiple accounts are involved: the supplier may send one bill covering several sites, while the utility continues to send a separate delivery bill to each site. Accounting may prefer that the supplier send a summary bill, consolidate its bill onto the utility's delivery bill, or follow a specified process (e.g., EDI) for submitting it electronically.

Leasing

If the client is a commercial real estate facility, Leasing (also called Tenant Relations) may need to inform tenants of a change in suppliers and electric rates. Attention may be needed to the wording of existing leases to avoid possible confusion. In one case, an enterprising sub-metered tenant found that his lease called for paying the landlord "the average rate calculated from the monthly bill from the utility." Once the landlord switched to a non-utility supplier, however, the *utility's* bill shrank when it no longer included supply charges. The tenant then claimed there was no requirement that it pay for charges from a *non-utility* supplier. To avoid this problem, only power suppliers whose charges can be consolidated onto the utility's bill may be chosen.

Public Relations and Environmental/Sustainability

Because customers may choose a supplier whose power comes from environmentally acceptable sources (e.g., wind) instead of utility-based sources (e.g., coal), the choice of supplier and its contract conditions may be of interest to those two offices. Choosing a supplier whose power comes from earth-friendly sources could help fulfill a corporate mission statement or commitment, polish a company's image, and/or reduce its carbon footprint. When an organization buys cleaner power, that's an ideal subject for a press release, its advertising, and featuring on its web site.

Marketing

If the customer has something that may be of value to a supplier (e.g., use of the customer's name and/or symbol in the supplier's advertising), there may be a basis for seeking a slightly lower rate in exchange for granting temporary use of such assets. An airline, for example, could offer frequent flyer miles to the supplier's new customers in exchange for an electric rate cut or better contract terms for the airline. A university could offer to rename its athletic field or a building for the term of the contract. A large retailer could agree to promote the supplier as its power source on printed receipts, or paychecks, or at its web site.

Failure to involve appropriate personnel and to consider their issues has caused trouble or lost opportunities for some customers. Newcomers to the process are urged to consider using a (at least the first time) to help them navigate such challenges and opportunities. For a discussion of the pros and cons of each type of specialist, see Appendix F.

Chapter 12

Options for
Competitive Supply Pricing

PRICE CONTENT AND STRUCTURE

The contents of a quoted supply price may depend on how the customer wants his price structured. Recall that, when competitively buying electricity, only the supply component (NOT any aspect of delivery) is involved.

All supply pricing should contain the components seen below (unless not appropriate to a state or utility). Some items may be excluded in a quote when buying floating retail pricing that is based on (i.e., "indexed" to) an ISO's wholesale market pricing. In such cases, only an adder (typically fixed) to that wholesale price is quoted that contains capacity, ancillary services, and other non-energy charges.

To avoid unpleasant surprises (and regardless of price structure), a customer should ask to see a sample invoice for his desired price structure to clarify such issues: contract wording alone may not be sufficiently clear. Line losses, for example, in some utility territories have been as high as 8%, which could make an otherwise seemingly sweet deal into a multi-year headache.

Energy is the raw kWh that a facility consumes, regardless of the generation source. For a fixed $/kWh contract, the price may be based on a weighted average (using stipulated monthly usages) of the forward wholesale prices quoted on the day the PPA is signed. Because no two facilities are identical in their monthly power usage or patterns, trying to compare pricing from two different sites is not a viable way to determine if one price is "better" than another. For a floating priced contract, the kWh price varies with the wholesale market, typically the hourly locational marginal price (LMP) for the customer's zone. The contract should be specific: e.g., Day Ahead Market (DAM) or Real-Time (RT) market? Based on actual hourly meter readings, or using simulated hourly readings based on the utility rate class profile? Energy may account for 70% or more of a fixed power price.

Congestion is charged by the ISO to the supplier when power is re-routed or secured from a more local generating source due to high load on the transmission system. While typically built into the ISO's zonal wholesale price, a few cases have occurred wherein a supplier tried to pass on unusual additional one-time congestion charges (e.g., during a period of severe weather), even for a fixed price that should have hedged such costs, but they have been rare.

Capacity ensures sufficient generation is available to meet a customer's annual (typically summer) peak demand (determined by the utility or ISO), except in Texas which, at the time of this writing, did not have a wholesale capacity charge. It may be based on a facility's peak demand at the time the entire grid peaked, or as an average of the facility's highest peak monthly demands seen in the summer months (as shown on the utility bills) regardless of the times they occurred. That kW quantity is then bumped up by a reserve margin (which may be 10% or more) to derive the customer's capacity "tag." The tag may be revised annually based on the peak seen in the prior 12 months, or during a different defined 12-month period. To calculate the cost, the tag kW is then multiplied by a capacity price in $/kW-month or $/MW-day, depending on the ISO. That pricing is developed via monthly, seasonal, annual (or longer) wholesale capacity auctions held by the ISO. Methods vary across the organized markets. Capacity may account for 4% to 15% of the total supply bill, depending on local conditions and a facility's load factor (a lower factor means a higher percent of the bill will be related to capacity).

Ancillary services are billed by the ISO to the supplier who passes it on to the customer. Those services include voltage support, blackstart, frequency regulation, generation reserves, and others. It may account for several percent of the total supply bill.

Other ISO or transmission charges related to delivering of power through the transmission system. While typically built into an ISO's zonal wholesale price, an occasional extra transitional charge has occurred as organized markets developed. For a time, a few PJM-based utilities levied a transmission charge (called "network integrated transmission service" [NITS]) on retail suppliers that were (in most other markets) instead included in the utilities' delivery bills. Occasionally, a federal or state regulatory change on how transmission charges are shared by ISOs and utilities may occur. Such charges may appear as line items added to the energy price.

Line loss is billed to the supplier by the utility. It covers energy lost in the utility's distribution system while delivering the supplier's power to the customer's meter. While varying among utilities (and even among suppliers), it may account for several percent of the total supply bill. It may also be called "Unaccounted For Energy" (UFE).

Renewable Portfolio Standard (RPS) is a charge based on a state-level requirement wherein a defined percentage of a supplier's electricity must come either from renewable sources or is supported through the purchase of equivalent Renewable Energy Credits (RECs). It may be built into a fixed price, or charged as a separate line item. It typically accounts for less than 1% of a supply bill.

Broker's commission (if the power was bought using a broker) may be ~1% to ~12% of the total supply bill, depending on kWh volume, total price, and broker greed.

Hedging is built into fixed pricing, and covers costs and upcharges incurred by the supplier when buying (e.g.) forward blocks of wholesale power at a fixed rate, and/or financial instruments (such as futures and options) to ensure a fixed price. Hedging may be 5% to 30% of the energy portion of a fixed price depending on the term of the contract and volatility of the wholesale power market. It may not be involved in floating pricing, since there is no price guarantee.

Overhead, management, and profit (sometimes called "OH&P") of the supplier may be 5% or less of the total supply bill, and would be built into a fixed energy or fixed adder charge. If the bill is consolidated onto the utility invoice (as described in Chapter 12), the extra cost of doing so would be part of the OH&P. Also built into that component is the commission to the supplier's account rep for handling the deal, which is not related to the broker's commission.

Taxes—typically the same rate as charged by the utility for supply, but open to the various options (e.g., exemption) discussed in Chapter 6.

After getting a grasp on the types of power products, we will look at ways to trim the cost of those supply price components.

COMMON TYPES OF POWER PRODUCTS

Various types of pricing, i.e., power "products" exist. Many suppliers provide short descriptions of their products at their web sites. Each is designed to balance a customer's tolerance for price risk with the desire for lower or stable pricing. Note that *price* risk is *NOT* supply risk: the lights don't go off if the price jumps (unless the bills are not paid). Instead, it means that there is a greater chance that monthly price may vary more than expected or desired, and total annual cost may be higher than budgeted.

Fixed

The option taken by many customers due to its simplicity, all components are fixed, making it essentially a hedge against a FAC or other variable utility pricing. As long as a customer's monthly usage remains within a defined variance relative to usage in the same month in the prior year (called "swing" or "bandwidth") that's what will be charged for supply. If significant changes have recently occurred at a facility, it is up to the customer (or the specialist) to instead provide projected monthly kWh usages that will then be stipulated in the contract. A typical swing is at least +/- 10%, with many suppliers offering +/- 25% (some offer 100% or unlimited swing). Any usage outside the swing may incur penalties in addition to the cost of securing or disposing of the differential kWh volume involved.

Unless a customer has very large process loads (e.g., industrial), or uses electricity for space heating, or has recently altered its facility, it would be unusual for the monthly kWh to vary more than 25% relative to the same month in the prior year. An identical or similar range may be provided for capacity based on a customer's annual peak kW load.

Collared

Total supply price is allowed to vary month to month, but within a price range stated in the contract. A variation on this method is "capped" pricing which won't rise above a defined level, but could drop lower, based on the wholesale energy market price.

Floating (also called "indexed")

A customer may specify that some (or all) of the previously listed components be allowed to float, while others remain fixed. The usual process is to allow pricing of the energy portion (i.e., the kWh) to float based on wholesale market pricing (i.e., the "index"). The other components are gathered into a fixed $/kWh adder that is constant from month to month. The varying price of a floating component (in this case, the energy) is shown as a separate line item on the invoice.

When all component prices are floating, the supplier's adder may represent just his overhead, management, and profit, and (if a broker is involved) the broker's commission. All (or just some months) of a floating price contract may be converted to fixed pricing with sufficient notice (stated in the contract). A customer may also mix fixed and floating pricing by stipulating which months fall into each option. Floating pricing raises the customer's price risk (relative to fixed pricing), but doing so generally yields the lowest average annual price over several years.

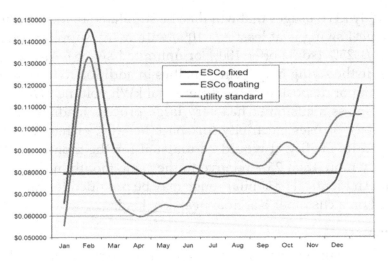

Figure 12-1. Retail commodity electricity pricing across a year. The fixed price for that term was just under $.08/kWh (horizontal line). The top line is the variable price from a non-utility supplier (ESCo), while the lower line is the utility's variable monthly price (including FAC). While the ESCo price is higher in winter and spring, it's lower during the summer and fall. Depending on a customer's monthly power consumption in each of those seasons, the annual cost for the ESCo may or may not be cheaper than remaining with the utility. *Credit: author*

Heat Rate

This is essentially a floating price where the index is the local wholesale price of natural gas instead of the local wholesale price of power. A heat rate (i.e., Btus of energy per kWh generated), is the overall efficiency of generation on which pricing may be based. The higher the heat rate, the higher the factor to be multiplied by the daily (or monthly) wholesale price of natural gas.

Block and Index

This is a mix of floating and fixed pricing wherein a customer may choose to buy a fixed-price block of energy (e.g., 60% of projected monthly kWh) and let the price of the remainder of the usage (the other 40%) float with the wholesale market (i.e., the index). A fixed adder to all kWh is included to cover other components. This option allows a customer to tailor the desired level of price risk, while not requiring the supplier to spend much on hedging, beyond ensuring the price of the block. Where smart metering is available, a block may instead be all kWh below a stated hourly kW load within a stated time frame (e.g., 7 AM to 11 PM), with floating pricing for all other usage above that kW level or outside that time frame. This process mimics how wholesale blocks of power are sold.

Guaranteed Savings

A rare option sometimes available when a power market first opens, this product offers pricing that is always a defined percentage (typically 5% or less) below the utility supply (not entire) price, calculated on a monthly average basis. Because a utility's price may vary (e.g., due to time of year or from a FAC), this product is effectively a floating price wherein the utility monthly supply price is the index. Such offerings tend to disappear as markets mature, but are an easy way to ensure some savings relative to taking supply from the utility. It's important, however, that the customer have some means to independently verify what the monthly utility supply price would have been for his facility. Some utilities provide calculators for their supply pricing at their websites for this purpose.

Contracts for Differences (CFD)

CFDs are more like a financial than a physical product. They offer an unhedged fixed price (which includes the supplier's adder) wherein the supplier and customer agree on a fixed $/kWh number, but the power is bought at wholesale, or from the incumbent utility. If the market power price (typically averaged monthly) is higher than the agreed-upon number, the supplier pays the customer to bring that price down to match the number. If the market power price is lower, the customer pays the supplier to bring that price up to match the number. Savings derive from the reduction in hedging costs. This, too, mimics a typical wholesale power pricing arrangement.

A CFD may work well when a customer has only a few

accounts. It can, however, be difficult to manage with many accounts because each account will have a separate stipulated price, based on its individual usage pattern, volume, etc. In one case where a customer had hundreds of accounts, this complexity greatly upset that firm's accounting department and CFD pricing was abandoned.

Various other (and more exotic) products exist, but the above reflects those most common. A good supplier (or specialist) will lay out these options for a customer and explain their pros and cons.

Chapter 13

Managing Competitive Power Procurement

One of the benefits of competition is that it offers customers choices, including different ways to select a power supplier. Getting the most out of the competitive market involves attention to that process. Various levels of professional help are available, and it makes sense to take advantage of it, especially the first time around. For details on options for such assistance, see appendix F.

FINDING AND CHOOSING RETAIL POWER SUPPLIERS

Dozens (in some case, over a hundred) licensed suppliers serve retail customers in each organized market. In the few deregulated states (e.g., Oregon) that are not part of such markets, only a few competitive suppliers may exist. The first step in choosing among them is to develop a list of bidders. A good specialist should be able to identify the 5 to 10 major players. Word-of-mouth recommendations from similar customers or their trade organizations may do the same. A more formal process (best for first time buyers and large organizations) may start with a request for information (RFI).

RFI CONTENT

Merely assuming that a supplier is acceptable because it is licensed may not be sufficient: many are essentially brokers willing to charge high pricing to unsuspecting (and/or ignorant) customers. When a contract worth a few hundred thousand dollars (or maybe a lot more) is involved, due diligence requires a professional approach.

To initiate the RFI process, develop a document (essentially an interview form) to send to suppliers found at your PUC web site that are listed to serve your customer rate class behind your utility. Your firm may already have a RFI template for vetting suppliers of other products that will help you get started. While a RFI could mean contacting dozens of suppliers, rest assured that only a small fraction are likely to reply. Many deal only with residential or small commercial customers, or may be unwilling to answer your questions. The RFI process allows you to tell your superiors that you gave everybody a fair shot (important for governmental and non-profit customers). Those that fail to respond to the RFI (or do so incompletely) are then justifiably dropped from the list. Following are the types of information many customers have requested from suppliers.

Relevant experience/resources:
- electric sales history in your zone: annual MWh and dollar volumes
- at least two customer references (in your ISO zone; if unsure which you're in, ask your utility)
- contact information (name, title, address, phone, email) for the rep that will service the contract

Financial/reliability indicators:
- supplier's type of business organization: e.g., corporation, LLC
- S&P/Moody/D&B or other standard credit rating (especially if contract term is longer than a year)
- book value of company (or its parent)
- is it (or its parent) publicly traded?
- list any markets entered and later dropped

Pricing/service offerings:
- list product options, e.g., fixed price, indexed, etc., or supplier's URL at which they are described

- is procurement the only service? List other energy service offerings (e.g., demand response, energy efficiency).

Only if you're interested in them, ask about:
- renewable power or REC offerings, Environmental Disclosure Label (EDL)*
- billing options (e.g., one bill for multiple accounts, summary, web-based, consolidated bill)
- financing of energy upgrade options (e.g., efficiency, on-site generation)

In addition to the filled out RFI form, also request:
- a standard fixed price supply contract and confirmation page in a word-searchable form (.pdf or .docx) showing all terms and conditions
- supplier's utility data access form letter** (only a few suppliers require use of their own form)
- indicative (i.e., non-committal) fixed price for 1 year for your largest account. Tell the supplier that the filled out data access letter will be sent upon your receipt of the supplier's filled-out RFI form. The form should include a statement to be signed by the supplier that any information you provide will be kept confidential and not used for any other purpose. The supplier may ask that you sign a similar statement regarding their quoted pricing. For those that respond, provide the account number, service address, utility data access letter, and a defined start date (e.g., the 15th of the next month). You may not get indicative pricing, but it's a handy way to get a hint of pricing for your load.

---

*An EDL shows, for the prior year, the percentage of electricity by energy source (e.g., coal, wind) as generated and/or purchased by the supplier. Customers may use that data to distinguish suppliers based on how carbon-heavy, clean, or renewable are their sources.

**To secure pricing from a supplier, it must access monthly usage and other data from your utility. A letter from your firm (on its letterhead), signed by you and listing the account number and service address (found on your utility bill), is needed for that purpose. Some suppliers require use of their own form, so ask for it in the RFI.

Review the RFI responses and sample contracts to whittle down the list of licensed suppliers to a dozen or less (you need at least 4 to ensure competitive pricing). Don't be surprised to find a few contracts with unacceptable provisos that, if a supplier is unwilling to change, will knock it out of the running.

Next, determine your preferred price discovery method.

PRICE DISCOVERY METHODS AND TOOLS

To squeeze the most savings out of the competitive procurement process, employ the best tools for doing so. Do not merely accept a price that a broker or supplier claims to be the "best available for your company." Without testing the marketplace, there is really no way to verify that statement and—in my years of experience—such general statements are often wrong.

Many purchasing professionals approach power procurement the same way they buy desks, cars, or computers. They develop a paper request for proposals (RFP) and put it out to bid. They examine bids at their leisure, assuming that they will remain good for 30 days because that was a condition for responding to the RFP. When they see an acceptable price, they review the purchasing contract from that bidder.

Following that outdated discovery process for electricity supply is fraught with potential problems. If seeking a fixed price for the term of the contract, a supplier is unlikely to hold his bid for 30 days: he may decline to hold it for more than 3 hours. Asking for a bid to be held for much longer may result in unnecessarily high bids when suppliers build in cushions to account for market volatility.

Because contractual issues may affect bid price, contracts should instead be reviewed *in advance of bidding* to ensure an apples-to-apples comparison occurs on the day of the bid. Some contract terms may impact pricing, so it pays to have all terms (e.g., term length, product type, swing allowance, etc.) comparable for all suppliers you expect to bid.

Once all contract issues are ironed out, experienced specialists may use online reverse auctions or tightly timed "blind" bids to elicit bids where only pricing (not contract terms) will vary. Because the discovery process must be detailed in the RFP (including bid dates and procedures), it's essential to make that choice in advance of issuing an RFP.

Some customers prefer *online auctions*. They provide an exciting and visible real-time process in which pricing drops as the customer watches on a computer screen. While popular for many years, some customers have found that such auctions may not be providing all the value available from the bidding process. Depending on the number and types of accounts, such auctions may limit account-specific pricing and price structure options.

Before agreeing to use an online reverse auction, ask to "sit in" on one (i.e., watch an actual auction for another customer) to get a sense of how they work. Be clear on these issues:

- Is it a visible on-screen real-time auction, or un-timed emailed pricing (a standard broker practice)?

- Are there any limits on pricing options, or account grouping/size/type? Auctions may require a simple response from bidders (e.g., bid only one account or group at a time) which may require aggregating accounts into groups, or by type or size, any of which may lose some of the benefit of competitive bidding.

- Who vets and selects the bidders? Does the customer get to see the list in advance and cull out any with whom he would rather not do business? Are any desired or major suppliers missing?

- How are auction 'savings' benchmarked? Some auctioneers create an opening bid with an absurdly high price to make it look like their service yields big savings. Prior to an auction, an experienced consultant should be able to provide a better estimate based on the wholesale forward price curve (discussed below) for the customer's zone, its monthly consumptions, local capacity pricing, and the customer's annual peak demand.

- Can the bid simultaneously include any add-options, e.g., XX% renewable energy, or does that require a separate bid and extra cost?

- What is the auctioneer's commission and with whom will it be shared?

- Will sample contracts of all bidders be available for review prior to bidding? Many suppliers' contracts contain provisos that, if not known prior to a bid, may be difficult to challenge thereafter (e.g., use of customer's name, limits on self-generation, nasty termination clause).

- If a customer doesn't like the winning bid, is it clear that he can just walk away from the process, with no negative repercussions? Some auction services charge a flat fee for each auction instead of taking a brokering commission. While that option may end up costing less, the customer signs an agreement to pay the fee, whether or not a bid is accepted.

Auctions can be fun (some CFOs love to watch them), but a 1 mill/kWh auction commission for a bid involving a large client (e.g., medical center) could exceed $50,000. For less than that, a customer may buy online auction software and run his own online pricing events. Such software could pay for itself in your first power auction, and then save when buying other products. Two of the major auction software vendors are Ariba (www.ariba.com) and Ion Wave Technologies (www.ionwave.net). When using customer-owned software for the first time, it pays to have an experienced consultant involved in the process to avoid embarrassing or costly mistakes.

An alternative to online auctions is the *tightly timed blind* (a/k/a "sealed") bid. It is aptly named because—unlike auctions—bidders cannot see the bids of their competitors, but the specialist and the customer can. Here's why that's important: when bidders can see all bids as they are made, each will bid just a bit lower than the leading bidder, but not as low as they might have if that information was not visible to them. The auction will eventually settle at a point when all bidders refuse to go lower (typically in less than an hour). As a result, the winning bid may still be higher than if competing bids had not been visible.

In a timed blind bid, all suppliers submit their emailed bids during a brief pre-arranged time window (e.g., 15 minutes). If the consultant running the event does not like the results, he may ask that the bidders re-submit bids below a target price lower than the lowest bid seen in the first round. That target may be set by prior discussion with the customer, or be a percentage below the lowest prior bid. If no second bids meet the mark, the session may be closed and re-run at another time, possibly with additional or other bidders.

Note that this process allows a customer (or his consultant) to approach his preferred supplier and tell him (even if he is already the lowest bidder) that he would win if he "just matched the lowest bid of $XX per kWh." Since the lowest bidder does not know that he is (remember it's a blind bid, not an auction), nothing (except maybe ethics) stops a customer from claiming he already has a lower bid in hand. If, however, that bidder fails to budge, the customer may have painted himself into a corner: does he take the preferred bidder's "higher" number (and perhaps look like a deceiving fool), re-bid the whole process, or what? Before taking such a step, carefully think through your strategy.

Here are actual $/kWh fixed prices from a blind bidding session involving seven suppliers:

A	$0.11960
B	$0.10896
C	$0.08070
D	$0.07983
E	$0.07911
F	$0.07897
G	$0.06620

If this had been an auction where all competitors saw each other's bids, G would likely have bid just below F (at maybe $.078), instead of $.0662. That would have cost the 6 MW customer an extra $400,000 per year! If the customer had instead chosen to simply remain with his incumbent supplier (bidder C) without any competition, he would have paid over half a million dollars extra.

Note the wide range among the bidders: bidder A asked for almost 80% more than G. Such price discovery tools may yield serious savings on the supply component of an electric rate. The consultant's fixed fee on this deal was (alas) only a small fraction of the savings.

This was an unusually clear case. I was never certain why G bid so low. It may have needed to quickly unload excess volume, perhaps due to a broken contract with another customer, or secured such power from another supplier looking to unload it for quick cash. In the end, a customer need not be concerned with the reason. As long as the winner honors the price for the term of the contract (which he did), the customer is the winner.

Reverse auctions were popular for years, but some of those promoting them have gradually shifted to timed blind bids for some of their pricing events. Careful preparation is needed to make them work properly: contracts (without pricing) must be reviewed (and, if necessary, negotiated) well in advance, bids must be standardized and coordinated, etc. A winning competitor's bid may be available for only a few hours, so a quick review and signing process must also be pre-arranged.

RFP CONTENT

Once the price discovery method and bid date are determined, it's time to create the electronic RFP. To avoid internal issues, start with a RFP previously issued by your firm, and adjust it as needed instead of starting from scratch. The RFP should contain:

• description of the desired price structure(s) (e.g., fixed, floating) to be bid

- accurate listings of all account numbers and service addresses, by utility, to be bid

- signed letter or form giving access to your utility billing data, listing all account numbers and service addresses to be covered in the bid, and any IDs or passwords needed to do so

- pricing spreadsheet to be filled in with both separate and aggregated pricing or adders in cents per kWh to at least 4 decimal places; if seeking floating pricing, also ask for fixed pricing (you may need it later as a benchmark). Doing so will avoid receiving responses in formats that may be difficult to quickly compare

- audited financial statement, or other credit data, for your company

- if tax-exempt, attach exemption certificate

- desired start date, taking into account meter read date and utility switching notice

- desired contract term options (e.g., xx, yy, zz months)

- which accounts desire or are capable of non-firm service, if any, and by how much kW

- any expected significant load changes by account (+/- X%/yr), or a forecast of monthly kWh/kW for each account that may experience such changes during the proposed contract term(s)

- desire for other options (e.g., demand response, savings verification, xx% green power)

- list customer assets that may be offered in trade for lower price (e.g., use of name/symbol)

- if required by your Purchasing Dept., your firm's pro forma purchasing contract or boilerplate text

- indicate choice of reverse auction or timed blind bid, with instructions for accessing the desired process (e.g., "email bids by noon, Tuesday, March 10, using the spreadsheet provided.")

To make sure all goes smoothly, follow these steps:

- just before issuing the RFP, verify the contact information for the supplier account reps that will respond to it; this industry has a lot of "churn" so whoever you spoke to a month ago may no longer be with the supplier, or may have been moved to another division. Sending the RFP to the wrong person is almost a guarantee you won't get a timely or appropriate response.

- in your email to each supplier with the RFP attached, request immediate acknowledgement to verify its receipt. If, for some reason, you have not yet reviewed its contract (shame on you), request that it be sent as an attachment to the acknowledgement.

- set the bid date at least 2 weeks after the expected date for emailing the RFP (longer if more than 4 contracts need to be reviewed, up to 4 weeks)

- note in the RFP that supplier Q&A concerning it is limited 1 week, and only via email (to maintain a proper paper trail).

To coordinate the chronology of those steps, see the graphic timeline in Figure 13-1.

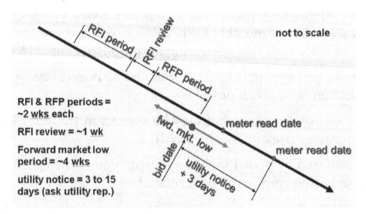

Figure 13-1. Timeline for planning competitive power procurement. Start from the period when the forward market is typically lowest for fixed pricing, (covered in Chapter 14) and work backward and forward to schedule RFI and RFP periods, and proposed contract start dates. *Credit: author*

Chapter 14

Ways to Save Money and/or Increase Value

Customers have many ways to reduce supply pricing by squeezing more value out of the competitive bidding process. In roughly chronological order, consider the following.

Minimize (or Avoid) a Brokering Commission

Brokers can be made to compete just like suppliers. Be ready, however, to verify the reduced commission with the suppliers being offered by the broker. Alternatively, use a consultant working under a fixed fee lower than a broker's commission (based on your projected kWh for the desired contract term). Alternatively, and with sufficient training, a customer's purchasing personnel may learn power procurement techniques and instead do the job in-house instead of paying a specialist. The Association of Energy Engineers offers several training programs on energy procurement.

Run Your Own Auction

As previously discussed, the commission paid to an online reverse auctioneer could be greater (for a large customer) than the cost of the software to do the job on your own. Assume a 1 mill/kWh commission (unless the auctioneer says otherwise), multiply by the projected kWh volume in the desired contract term, and see if buying the software may be cheaper, either in the first or subsequent online auctions.

Aggregate/Separate Accounts

A customer with multiple accounts may bring together (i.e. aggregate) all power supply needs under one contract, expanding the total kWh volume to be purchased. A single fixed price (or fixed adder) covering all accounts is the simplest option, but may leave money on the table if not done properly. Smart buyers employ two additional methods. To make it easy, let's assume we have 5 accounts and 2 suppliers (see 14-1).

Example (energy only)		
Acct #	Splr A	Splr B
0001	$.044	$.040
0002	$.049	$.051
0003	$.069	$.063
0004	$.046	$.042
0005	$.052	$.055
agg. price $.052		$.048
aggregated savings = 7.7%		
savings by ind. acct. = 9.3%		
Savings increased by 20%		

Figure 14-1. With multiple accounts, always ask for separate and aggregated pricing in order to "mix and match" best pricing by account, and to create leverage to seek lower pricing from one bidder. In this example, securing lowest individual pricing from two suppliers saves more than accepting an aggregated pricing from one of them. *Credit: author*

- Always ask for both separate *and* aggregated pricing. As seen in Figure 14-1, Supplier A is cheaper than Supplier B on 2 accounts, and the reverse is true for 3 accounts. This may occur because suppliers have differing ways to price each account. It may then be possible to push Supplier B to drop his price on the 2 accounts to equal (or beat) Supplier A's pricing. If B refuses, split the accounts between the two suppliers and sign a contract with each to get the lowest overall cost. In many cases, I have seen sufficient wiggle room for B to come down to match A, so all 5 accounts can then be awarded under one contract, with separate pricing by account, or through a single fixed price for all accounts that yields the same total cost.

- Nothing says you must bid out all accounts to a non-utility supplier. Be ready to triage some of your accounts by leaving any having an individual price (determined by getting separate pricing per account) higher than the projected utility supply charge for those accounts. Consider the utility to be just another supplier with a possibly better price for some of your accounts. Leave the "bad" ones with the utility to be cross-subsidized by other customers in the same rate class still taking supply from the utility. A quick way to gauge in advance which accounts may be priced higher than others is to calculate their annual load factors (LF). As seen in Figure 14-2, those with low LF are more likely to be priced higher. Prior to bidding, determine their likely utility supply price under the tariff plus historical FACs (or the standard offer, if available) so they may be quickly dropped from the bid stack if prices from competing suppliers exceed it. In the diagram, the circled account's fixed price quote was higher than the utility's standard offer (the dashed line) and was thus left with the utility.

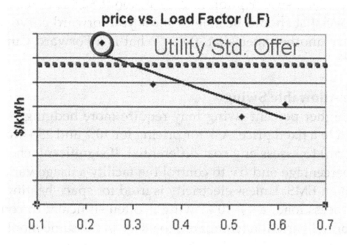

Figure 14-2. Power pricing for accounts with low annual load factor (x-axis) may be higher from a non-utility supplier than from a utility. Pricing for four accounts shows that three are below the utility standard offer, while one is higher. Leave it with the utility and take competitive pricing for the other three. *Credit: author*

Get pricing for different term lengths

A longer term results in a larger contracted kWh volume. That may draw more bidders, but—depending on the forward curve (explained below)—may result in a higher or lower fixed price due to projected wholesale market conditions. A longer term (e.g., 3 years) may also involve extra hedging costs that are passed on to the customer. To test the market, request (in the same bid) fixed pricing for several terms (e.g., 1-, 2-, and 3-year).

When the forward price curve is dropping, it may make sense to lock in a longer term to capture that lower projected wholesale pricing. If the reverse is true, it may make more sense to choose a short term (e.g., 1 year) and then re-check the forward curve later to see if prices down the road get better or are at least stable. A good source for forward curves for many wholesale markets across the U.S. is *Megawatt Daily*, a Platts. com publication (www.platts.com/products/megawatt-daily). It is not free; a subscription in 2017 is about $2,500 a year. Any good specialist should be able to show you forward curve data. If not, try another specialist. (See "What is a Forward Curve?" opposite.)

Reduce Allowable Swing

A higher percent swing may require more hedging, which may add to a fixed price. Ask for pricing for 10% and 25% (for the same term) to assess any cost differential. If significant, choose a lower percentage and try to control the facility's usage variation through an EMS. Unless electricity is used for space heating or a large process load, a +/-10% swing is often sufficient to contain most normal variation (compared to usage in the same months in the prior year). A 25% swing may do so without a need for controlling usage variation.

Avoid a Deposit or Consolidated Billing Charge

When a customer's credit or utility payment track record is poor (e.g., many late payments), some suppliers request a deposit equal to one month of power supply. While refundable at the end of a contract (assuming payments have been on time), the time value of the money locked up by the supplier may be significant. Alternatively, a supplier may quote a high late payment penalty

What is a Forward Curve?

Wholesale power is sold through various private exchanges (e.g., the Intercontinental Exchange, a/k/a ICE). Platts and other sources monitor their trades and offers to daily derive monthly pricing for future months for on-peak (i.e., 7 AM to 11 PM weekdays) and off-peak (all other hours) block pricing were it to be bought today. Figure 14-3 shows a graph of those pricings, which is called the "forward curve." When future pricing is rising (across years, not just month-to-month), that trend is called "contango." When the curve is dropping, we call it "
." Good specialists watch the forward curve on a daily basis for markets in which they are active to find buying opportunities. Note that the forward curve is only for the energy (kWh) portion of the supply bill. It does NOT include capacity, transmission (unless quoted by zone), delivery, tax, etc.

NORTHEAST PLATTS M2MS FORWARD CURVE: ON-PEAK

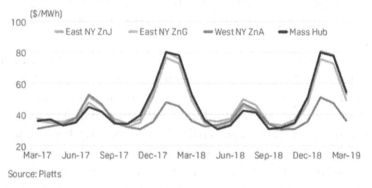

Source: Platts

Figure 14-3. Forward on-peak (7 AM-11 PM) pricing for wholesale energy-only (capacity and other components excluded) for four zones, as seen from February 2017. *Credit: Platts*

(i.e., over 1.5% per month). In both cases, an alternative may be a letter of credit (LoC) from the customer's bank to the supplier. If the customer fails to pay on time, the bank instead pays the supplier, and the customer pays back the bank under a short-term loan at a lower interest rate.

An alternative method that avoids a deposit to the supplier and other credit-related issues is to request a consolidated bill.

Regardless of the pricing structure, the charge for supply then appears on your utility delivery bill as a separate line item. The utility pays the supplier, and you pay the utility for supply at the same time you pay it for delivery. Some suppliers do not offer consolidated billing, and those that do may charge slightly more since they must pay the utility a fee for this service. In essence, the utility becomes the supplier's collection agency. This process is called "Purchase of Receivables" (POR). If no credit issues exist, *avoiding* consolidated billing (which is the default for some suppliers) may yield a slightly lower price.

Manage More Price Risk

Each time a customer wants to fix the price of a component, a supplier may need to hedge it (i.e., take cost-fixing actions), adding that expense to his price. The end result is that a fixed price may be 10% to 20% higher (on average) than a floating energy price across a year. Some have claimed differences as high as 30%.

Note that does not mean a lower average floating price *every* year. In my experience, it has been true about 3 out of 4 years, with 1 of those years showing a higher annual cost under a floating price contract, typically due to extreme weather (e.g., a very hot summer). In a few cases (2011 Texas drought), that 1 bad year essentially "ate" most of the savings from the prior 3 good years. And nothing says you can't have two "4th years" in a row.

To adjust the level of acceptable risk, a customer may opt for a mix of fixed and floating pricing, as described earlier under "Common Types of Power Products." A portion (based on time, volume, months, or other characteristics) may be fixed while the rest floats with the wholesale market price.

In some markets, there may be two choices for a floating wholesale price. Many suppliers offer floating pricing indexed it to the day-ahead market (DAM) price, which offers hourly pricing for the next 24 hours. Some markets also offer a real time/hour-ahead market (RT or HAM) that gives pricing only an hour ahead. The latter may yield more price volatility, especially during brief weather extremes, but a slightly lower average price than the DAM. Deep knowledge of a market is needed to project that difference across a contract's term. In Figure 14-4, we see a comparison (for February 2017) between the DAM and HAM price options on a daily on-peak basis in several northeastern markets. A positive number means the DAM was more expensive than the HAM, while a negative number means the DAM was cheaper on the indicated days.

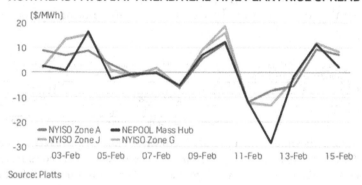

Figure 14-4. Wholesale on-peak energy-only power pricing for day-ahead and real-time (i.e., hour-ahead) purchasing for a two-week period in February 2017 for four zones. When the curve is above zero, day-ahead was more expensive than real-time. When below zero, day-ahead was cheaper, which typically occurs when weather conditions become extreme (i.e., much hotter or colder). *Credit: Platts*

Some of the risk from a floating price may be handled by maintaining a financial reserve in a facility's electricity account to cover months when the price is higher than expected. Alternatively, a letter-of-credit (LoC) with the customer's bank may be used to cover occasionally high bills.

For customers having multiple accounts, price volatility across years may be addressed through "laddering" of separate fixed price contracts, i.e., starting them at different times and/ or with different terms. As seen in the simple example in Figure 14-5, 4 accounts are started at the same time, but each ends at a different time. Each has a different fixed price. The impact of year-to-year market fluctuations is dampened (i.e., average pricing for all contracts together varies much less than that of individual contracts) because the contracts do not all end at the same time. Those who invest in the stock market may recognize this process as similar to "dollar cost averaging" under which an investor buys a fixed dollar amount of a stock on a schedule, regardless of its price at the time. In so doing, he will be purchasing more shares when the price is low and fewer when the price is high. The average price he pays won't be the best, but also won't be the worst. In our case, replace share price with kWh price.

	6 months	6 months	6 months	6 months	6 months	6 months	6 months
25% load	Ctrct. A: $.05	Ctrct. A1: $.07			Ctrct. A2: $.08		
25% load	Ctrct. B: $.06		Ctrct. B1: $.065			Ctrct. B2: $.045	
25% load	Ctrct. C: $.07			Ctrct. C1: $.055			Ctrct. C2: $.035
25% load	Ctrct. D: $.045				Ctrct. D1: $.075		
ave. price:	$0.05625	$0.06125	$0.06250	$0.05875	$0.06875	$0.06375	$0.05875

Figure 14-5. This simple example of "laddering" shows four fixed-priced accounts (A-D) in which all start at the same time but have initial contract terms designed to end 6 months apart. Note that, while pricing for individual contracts may vary greatly, the <u>average</u> power price (bottom line) for all accounts together varies only slightly. This process helps limit overall cost volatility for the accounts as a group. *Credit: author*

A customer may also start with a fully floating price and, over the contract's term, switch one or more months (or portions of his load) to a fixed price whenever the forward curve for those months looks advantageous. His price risk is gradually reduced as he fixes the price of more of his load. In doing so, he has avoided inadvertently buying all his power at a too-high fixed price. His floating price contract must, however, stipulate how the conversion from floating-to-fixed will be done expeditiously, and the notification time required before the trade will actually be made (which could be a few days to two weeks).

The task of picking out the best times to switch from floating-to-fixed may be given to the supplier, but doing so involves a significant amount of trust. While many suppliers tout their ability to realize significant savings for their clients, it has been my experience that some have done no better than the average investment advisor does with stock purchasing. If a customer permits his supplier to perform that task, the PPA (or a separate service contract) must spell out how any supplier commissions or service fees will be calculated, and how the customer may later withdraw his permission.

Fix Energy but Float Capacity

Suppliers may purchase wholesale capacity directly from a generator or buy it from the ISO (or a mix of the two). The $/kW-month price will likely be close to the capacity auction pricing developed by the ISO. But the customer's summer peak demand (on which his kW capacity tag will be determined) may vary from year-to-year, especially if he is developing demand response capabilities.

Suppose that a customer has a three-year contract. If he locks in a fully fixed price across all 3 years, but—during the first year—installs variable speed drives (VSD) on his air handlers. If he uses them to routinely limit his peak demand, he may end up reducing the peak he creates at the time that the ISO peaks. His kW capacity tag for the second and third years of his contract will then be lower, but the price he paid for capacity as part of his fixed price will not also drop because it was locked in when the contract was signed.

If such demand control capability may be installed during a multi-year contract term, it would be logical to float the cost of capacity. Some risk is entailed, however, because that means a jump in the ISO capacity auction price (e.g., due to retirement of a power plant) would not have been hedged, and will be passed on to the customer. On the other hand, capacity pricing may drop if a new transmission line into the zone is installed, bringing competitively priced capacity into the market. A good specialist should inform that decision.

Any routine reductions in monthly peak kW demand from such efforts will, of course, reduce monthly utility delivery demand charges.

Set Bid Date Based on Market Timing

In some markets, fixed forward pricing in the wholesale market bottoms out at about the same time each year (though that is never guaranteed). Figure 14-6 shows how forward pricing for various months (or groups of months, called "packages") bottomed out in the same month (March) as those months approached. When buying fixed pricing for a year, savings may be possible (relative to purchasing at other times) by setting a contract's price during such a dip. It may then make sense to set the contract's *end* date one month *after* such a dip so that—a month before the contract's end—bidding for the next year takes place while the market is "down."

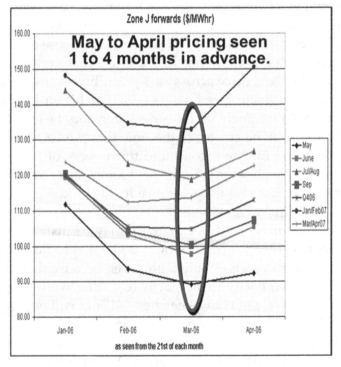

Figure 14-6. A study of fixed forward pricing for several months (or groups of months) showed that it tended to bottom out over time. Bidding may then be optimized at the point where most forward months are lowest (in this case, it was during March). This characteristic differs among markets and zones and is never guaranteed. *Credit: author, using forward curve data from Platts.*

I have often heard suppliers (and brokers) caution against waiting to "time" the market in this fashion because it is always possible that a better price may never be available later. In my experience, however, much of that talk is designed to get you to sign with them now to avoid another supplier (or broker) later persuading you to sign with them.

To qualify this option, a good specialist should be able to show you trends in forward curves for your zone. When done routinely from year-to-year, I have found that positioning fixed price contract start and end dates based on the usual times of such dips (which may follow natural gas pricing) is a good bet. It is almost certainly better than signing up in the dead of winter or the height of the summer, when near-term (and forward) power pricing is almost always high.

Offer Interruptibility Rights

The foundation of demand response (DR) is the ability to secure load reductions at the time that the ISO is about to incur very high hourly wholesale pricing. Note the "hockey stick" hourly pricing seen in Figure 14-7. DR participants may be paid handsomely for the right to request that they reduce load and/or start on-site generators that will minimize grid load at such times.

When a high hourly grid price is sidestepped, all buyers—suppliers, utilities, and floating price customers—will save.

Note, however, that DR calls may not happen, or be rare. What if the supplier could request a demand reduction whenever it was advantageous for both it and the customer, regardless of calls from the ISO? In essence, this is the same tactic as is used for interruptible gas service: a gas supplier has the option to ask a customer to temporarily switch to a stored alternate fuel (e.g., fuel oil). The gas supplier then instead sells the gas to the wholesale market that he would have sent to the customer, or sells it to other customers taking a floating price, but lacking interruptibility. The supplier then splits the profit with the interrupted client (as per the contract's terms). In this example, replace "gas" with "reduced kWh."

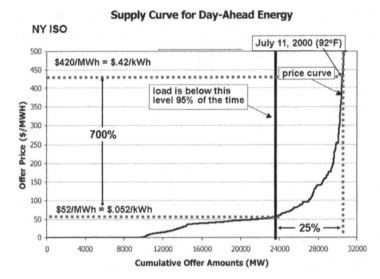

Figure 14-7. Wholesale hourly power pricing may rise exponentially as grid load approaches grid capacity. Demand response action during hours of extreme pricing may reduce load sufficiently to yield a significant drop in hourly pricing. *Credit: NY ISO*

Note that federal emission regulations may bar use of an on-site fossil-fueled generator unless done under the direction of an ISO during a DR call, or a utility outage. As power storage with batteries (which create no emissions) becomes economically feasible, it could instead be arbitraged to profit from interruptibility.

Mix in Demand Response Service

Many suppliers also provide DR services. Sometimes combining it with their power contract may elicit a lower DR percentage to be taken by the supplier, in effect putting more money in the customer's pocket than if the two services were contracted to two different companies.

Compete Line Loss Percentage

When taking a floating price, line loss may either be built into the fixed adder, or charged as a separate line item. When mixed into the adder, the supplier offering the lowest adder will win a competitive bid. If it is to be charged as a separate line item (remember I told you to review a sample invoice so that all charges would be clear?), it pays to know in advance the line loss percent that will be charged. The utility charges the supplier for that line loss, so you may think that it would be the same for all suppliers serving a given zone. Depending on the location of the generation used by the supplier, and which transmission line(s) are used to bring power into a zone, line loss percent may, however, vary. In some cases, I have seen it twice as high from one supplier as another, with the highest number being the line loss number quoted by the utility to its PUC. When seeking the lowest possible cost for electricity, a few percent could be the difference among suppliers. If two adders from different suppliers are very close to the same, choose the one with the lower line loss percentage. Note that this requires that your bid response form request line loss percentage along with the $/kWh price of the adder.

Trade Something

As discussed in Chapter 11, a customer may own a commodity that a supplier values, such as a well-known brand name. A customer may be willing to allow use of its name and/or symbol in the supplier's advertising or web site as a way to attract new customers. Co-marketing (e.g., an airline offering frequent flyer miles to a supplier's customers) may be another way to negotiate a lower rate, or a one-time payment to use a name or symbol during a contract's term.

Fold in Upgrade Financing

One of the major difficulties faced by institutional facility managers trying to implement energy efficiency projects is the competition for project capital. One way around that roadblock is to use the operating budget as a financing tool. Some energy suppliers are also energy service companies (ESCos) that design, finance, and install such projects. By building into a long-term (e.g., 5-year) supply contract a monthly fee that capitalizes a project, the extra cost may be small enough to fit within the facility's energy budget. During the contract's term, savings from both the competitive energy procurement and the efficiency project help counter that adder. To keep the accounting department happy, the project financing could be configured as an operating lease. At its end, the project becomes the property of the facility.

Generate On-site

A variation on the upgrade financing may be to have the supplier/ESCo finance, install, and operate generators at a facility (e.g., a CHP system or solar PV). It sells the output to the facility at a reduced rate, and the capacity into demand response programs. Depending on the type of generation, the system may provide emergency or backup power during an outage. That may add value to a facility while providing resiliency that many facilities desire, but cannot afford. For more on the pros and cons of on-site generation, see appendix G.

Cheaper Carbon Neutrality

Competitive power procurement allows a customer to specify a percent of its power to come from renewable energy sources (e.g., wind), usually at a premium (which may range widely, depending on location and existing subsidies). For general guidance on buying renewable power, go to: www.wri.org/publication/corporate-guide-to-green-power-markets

Unless a PPA is with an actual renewable power producer (e.g., an on-site PV system owned by a third party or a wheeling agreement with a remote PV system), a typical retail supplier may make his power "green" by instead purchasing renewable energy credits (RECs). Each credit

Figure 14-8. Wind turbines may offer low-cost renewable power that may be purchased as part of a Power Purchasing Agreement (PPA). *Photo credit: US DOE*

ensures a MWh of renewable power enters the grid to displace a non-renewable MWh. In the end, it's all the same atmosphere, so that action has the same climate impact as if the facility's actual power was renewable.

While a REC ensures use of renewable power somewhere, its actual carbon impact is based on the carbon content of the grid power that it displaces. In renewable-rich California, for example, a REC generated in that state avoids only about 500 pounds of carbon, while a REC from a wind farm in Wyoming (where most power comes from burning coal) may avoid about 2,100 pounds. As a result, fewer Wyoming RECs may be needed to counter the carbon created by a facility's electric consumption in a state having "cleaner" generation. A regulated utility may be required to buy RECs generated in its home state to meet a RPS requirement, but a non-utility supplier may have the latitude to purchase them elsewhere. So-called "national" RECs may cost only a fraction of local RECs, allowing it to offer cheaper renewable power to its customers. If considering use of RECs to reduce your firm's carbon footprint, ask for pricing both "green" and "brown" power to assess the differential. If more than a few mills, it may make more sense to buy RECs on your own. RECs are available through power suppliers and several trading exchanges.

If your goal is a reduced carbon footprint (and not a desire to subsidize the renewable power industry via RECs), consider securing carbon offsets as a potentially cheaper alternative. Each carbon offset ensures that a ton of carbon is being avoided or captured, typically through capping landfills (which blocks seepage of methane, a potent greenhouse gas) or planting trees (which absorb carbon from the atmosphere). For a given carbon impact, carbon offsets may cost less than renewable power. Each is traded through a different private market, so the value differential may vary over time. For general information on offsets and how they work, see http://apps3.eere.energy.gov/greenpower/markets/carbon.shtml?page=0.

In one case, a large hotel that touted its sustainability was buying RECs through a PPA to negate the carbon impact of its electricity use, and carbon offsets to do the same for natural gas consumed by its boilers. By instead using only carbon offsets to zero out its carbon footprint, the customer saved over $40,000 a year. Offsets may be purchased with or without involvement of a power supplier.

Few of the above options are available when buying power from a utility. To get the most value out of competitive power procurement, give some thought to each before configuring your RFP. Use the process to test the market for those of interest to your company.

Chapter 15

Negotiating/Improving
A Supply Contract

One of the options inherent in competitive retail power procurement is the ability to negotiate the supply contract. This opportunity cuts both ways:

- opportunities may exist for improving terms and conditions, and for squeezing out extra financial value

- involvement of an attorney (especially one being paid hourly) could significantly slow or hamper the process, while burning resources (your time and budget) with minimal dollar benefit.

The North American Energy Standards Board (NAESB) offers a standardized power contract that, while overly long and detailed, is a good and unbiased template. While it's always easier to start with a supplier's standard contract and propose ways to make it better, the NAESB document provides wording and direction that is generally accepted in the industry. An order form to purchase ($250 to non-members in 2017) and download a copy may be found at *https://www.naesb.org//pdf/ordrform.pdf.* Look for Standard Number RXQ.6.5. The document is not available in paper form.

Require that sample contracts be supplied weeks in advance of the bid date in electronic formats (e.g., Word, PDF). To find potential problems, word search for: default, security, termination, date, amount, name, court, automatic, change, renewal, late, fee, majeure, term, charge, loss, notice, dispute, payment, percent (or %), agent, assign, balancing, collateral, rate, capacity, generation.

Where possible, use an attorney with experience in utility regulation or law and (if possible) natural gas procurement contracts. Be clear that buying power is a price, not a contractual, opportunity. Seeking ideal legal wording won't yield better pricing. It could even increase price bids. The goal is to live with as much existing language as possible. Challenge only those terms that may create problems for your firm, e.g., an unacceptable indemnification clause.

Press hard for full concurrence on any wording changes no later than a week before the bid. If the attorney demands (e.g.) two months to review sample contracts, then schedule the bid date appropriately. Once bids are announced, only a few hours may be available for post-bid discussion before signing. Fussing over wording at that point could kill the deal, requiring a repeat of the RFP process.

While the attorney may focus on the "what ifs" in a contract, his lack of familiarity with PPA lingo may cause him to miss some important issues. I have also watched as attorneys too proud to admit ignorance of energy procurement terminology missed a variety of important issues. They instead obsessed over items with which they were familiar, but had little bearing on a typical fixed price contract's risks. To avoid such blunders, be sure to review and resolve the following pertinent issues.

- Are any of these price components not specifically included: capacity, line loss, ISO charges?

- Will the contract automatically renew at the end of its term? If so, how may that possibility be reversed or managed?

- Is the allowable swing unquantified or below 10%?

- Is there a "Material Change" or "Material Deviation" clause? If so, must the change be at least 25% before being invoked, and is it limited to non-weather variations?

- Is the *force majeure* "not limited," i.e., possibly allowing default on a fixed price unless price is excluded under that or another clause?

- Is the termination penalty based on something other than "liquidated damages," i.e., includes only those costs needed to make the supplier whole if a customer terminates the contract?

- Are any extra unrequested services included, e.g., a "free" energy audit/survey?

- Is there any objectionable wording, e.g., indemnification?

- Is the late payment interest rate greater than 1.5% per month (i.e., what utilities charge)?

- Is the payment term less than 20 days?

- Are there any requirements for new metering or equipment?

- Are there any technical restrictions on facility operations, e.g., blocking installation of on-site generation, or participation in demand response programs?

- How is balancing to handle an excess variance to be handled, and how are penalties quantified?

- Is use of the customer's name and/or symbol allowed in supplier promotions?

- If pricing is to float, how is the index defined, the bill to be calculated, and how may the index be verified (e.g., through a trade or ISO publication)? To clarify, request a sample invoice.

- May a deposit be required, either initially or as result of a customer credit issue? If so, may it be replaced by a letter of credit from the customer's bank or avoided through consolidated billing?

HOW TO IMPROVE A CONTRACT

Aside from fixing the problems listed above, the contract negotiation process may be used to seek opportunities having financial value. Each of the following has been granted by at least one supplier to at least one client.

- Allow one-time penalty-free late payment if made within 30 days (instead of the contractual 20)—some golfers call this type of latitude a "mulligan."

- If desired by your accountants, receive invoices via EDI. This is most appropriate when the number of accounts, and thus monthly invoices, is high (e.g., dozens).

- Seek an incentive (e.g., half a mill per kWh) for payment within 10 days via EFT. This may be most appropriate if/ when inflation again exceeds 7% per year.

- If dealing with an existing supplier seeking renewal of his contract, ask forgiveness of outstanding late fees as a condition to allow it to bid. I once got a supplier to give up $250,000 in such fees in order to be allowed to bid.

- If you want to stay with the existing supplier, seek a favorable "blend and extend"* (B&E) to remain with him instead of seeking competitive bids.

- Provide a one-time bonus for use of the customer name and/ or symbol in supplier promotions. I once secured a $100,000 payment for a customer to allow use of its name during a contract's term.

- Provide a one-time bonus for a guarantee that any new accounts opened during the contract term would go to the supplier. This only makes sense if there is an expectation of opening many accounts (e.g., a rapidly expanding chain store) during that term.

- Provide a one-time bonus for the right of first refusal to match a competitor's price for any new accounts opened (this is mutually exclusive to the item in the prior paragraph).

---

*Before a fixed price contract runs out, extend its term and capture expected low future pricing if the forward curve is backwardated (i.e., falling). That lower-priced future power is then mixed with the higher-priced power in the existing contract to immediately bring its price down via the averaging.

And don't be afraid to get creative: what else might you or your associates want in a contract?

While settling such issues before the bid date is preferably, a few could be discussed shortly after bids are in. If two bidders offer the same price and neither will budge or offer additional perks to seal the deal, a supplier may instead be flexible and accept one or more of the above to do so. Another way to break a deadlock may be to request:

- a higher allowable swing percentage (without an increase in price).

- XX% of the power to be made renewable via RECs (or increase a previously requested percentage) at no extra cost to the customer.

Inherent in every price is a commission to the supplier's sales rep (i.e., not a broker). If that person is sufficiently motivated to make the sale, there's always a few tenths of a mill (maybe more) that could be squeezed from his commission to make it happen. In addition to the above options, be prepared with several others ready for discussion during the 2- to 3-hour period between bids and the signing deadline. Have the costs of desired items in hand before making a request. A viable target is something worth up to roughly .3 mill/kWh multiplied by the contract kWh volume. Here are a few examples:

- if you're a non-profit, a tax-deductible donation, e.g., for computer or lab equipment, or a cash grant.

- wholesale pricing service or newsletter (especially handy to see forward curves and market pricing). Megawatt Daily costs about $2,500 a year, so (at .3 mills/kWh), an annual contract volume of about 8 million kWh (roughly a 2-3 MW combined peak load for all accounts) may be sufficient to support it.

- interval meters and/or sub-metering; assume $2,000 per metering point (or $500 per single strap-on metering device, including connection to a facility PC).

- energy accounting/management software: e.g., Metrix, EnergyCAP™ (have a quote in hand for the brand and version you like).

- energy audit for your largest facility (specify ASHRAE Level II but be ready to accept a Level I); a Level I audit is worth roughly $.05 to $.10 a square foot.

- greenhouse gas (GHG) inventory (using your preferred carbon footprint calculator/procedure).

- engineering evaluation of a lighting upgrade, voltage upgrade, PV system, or CHP (pick one).

- billing audit and/or delivery tariff review.

Don't feel constrained by that list. Before going out to bid, consider what small perks your facility might like, if it had the opportunity to request them. Determine their cost and how to secure them so they may be placed on the table for quick negotiation. But don't promise anything to the boss in advance: there are no guarantees you'll get anything beyond a good price quote.

POST BID OPTIONS

At this point you've secured pricing and chosen your supplier. If a fixed price was taken, there is very little oversight needed. If a floating price (either fully or partially) is involved, it needs to be checked monthly against whatever index (e.g., weighted monthly day-ahead ISO) is specified in the contract. One of the main "errors" found in bills with floating pricing is that, after a price spike has passed, the price somehow fails to drop back to a normal level. Bill auditors catching that error are very happy to take XX% of the easy refunds they secure for such customers.

If a floating price contract allows the supplier to lock in tranches of energy as low-price opportunities arise, be sure they do not extend the ultimate term of the contract without your permission. Otherwise, you may be unable to switch suppliers once the initial term has run out. In order to properly balance usage versus supply, utilities allow only one supplier to serve an account at any point in time, so you may need to wait until the last tranche runs out before you can again go out to bid.

While a good supplier will routinely inform a customer of opportune B&E options, some may not be so conscientious. It therefore pays to keep an eye (either your own or a specialist's) on the forward curve to determine if and when a B&E may be worth pursuing.

CONCLUSIONS

Nothing requires electricity customers to switch power suppliers, though that may change in the future. Nothing guarantees savings (relative to utility pricing) when buying from a non-utility supplier, unless the contract specifically says so. Retail power procurement, like many other options available in the commercial world, comes with both opportunities and pitfalls. Customers may pursue the process on their own, or use a variety of tools and/or specialists to assist them. The time and effort spent on doing so will help a facility cut or contain its costs for electricity.

Appendices

A. Helpful Links

Access to all PUC web sites:
www.naruc.org/commissions.cfm

Links to most electric utility web sites:
www.utilityconnection.com/page2b.asp

RELEVANT AEE WEBINARS

- load profile analysis, smart meters, demand response: www.aeeprograms.com/realtime/LPonline/

- buying power in competitive retail markets: www.aeeprograms.com/realtime/PowerPurchasing/

- small scale cogeneration: www.aeeprograms.com/realtime/CHP/

B. List of Utility Bill Auditing Companies

Listing does not imply endorsement, good service, or results. Most auditing firms serve customers behind only a few utilities or in just a few states. Those that serve nationally tend to cost more. Customers with facilities in multiple states may economize by using several small auditors instead of one large firm. Read contracts thoroughly before signing, and consider using a RFP to secure lower pricing and better terms from billing auditors.

www.nusconsulting.com

www.utilisave.com/
www.energysolve.com
www.troybanks.com/
www.utilityreduction.com/
www.ericryan.com/
www.utilmanagement.com/
www.utilitech.com
www.emr-energy.com/#panel-4
www.americanutilityconsultants.com/
www.nuenergen.com
www.summitenergy.com
www.utilityauditsolutions.com/
www.nationalpowersource.com/utilityauditing.html
www.utilitycheck.com
www.ecssaves.com
www.richardsenergy.com
www.naturalgas-electric.com
www.avalonenergyconsultants.com
www.utilitiesanalyses.com
www.abraxasenergy.com
www.monarchcostconsultants.com
www.energyconsultants.org
www.ucmc-usa.com
www.tdi-consulting.com
www.ausaconsultants.com
www.utilitydoctor.net
www.utilityauditing.us/
http://utilitybillaudittraining.com/
http://www.commercialutility.com/index.php
http://www.procurianenergy.com
www.auditek.net

To update this list, Google on "utility bill auditing service" and include your state's name in the search field. Doing so may net other nearby auditing services.

AN AUDITOR'S SCOPE OF WORK

Some auditors pursue only easy ways to cut costs, e.g., find billing and utility rate errors. To ensure a more robust level of service, consider this sample scope of work:

• benchmark usage, peak demand, and year-to-year consistency for the same months

• verify suitability of existing and alternative tariff rate(s) and taxes

• review monthly bills, find errors, seek refunds

• assess options for economic development opportunities and special deals

• evaluate and pursue deferral options

• ballpark viability of energy upgrades (e.g., lighting, PV)

• if interval metering data exists, examine load profile(s) for obvious problems or operating issues.

C. Understanding and Securing Interval Metering and Data

CONTROLLING PEAK DEMAND VIA DATA VISUALIZATION

Your facility's monthly electric bill arrived today and your peak demand was 1200 kW. During the same month last year, it was only about 1000 kW. That 20% jump raises a batch of questions:

• What happened?

• What time and date did it happen?

• How often, and for how long, did you approach that peak?

- What could have been done to avoid this problem?

- Could you save money under a time-of-use (TOU) or real time pricing (RTP) rate?

- Might you benefit from participating in a demand response (DR) program (i.e., reducing load or turning on backup generators when called upon to do so by the utility or ISO)?

If you (or your boss) have ever asked such questions, then it's time to start looking at your interval data and load profiles.

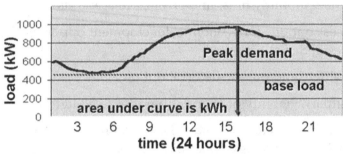

Figure C-1. Typical two-dimensional (power vs. time) weekday load profile for a commercial building, using quarter-hourly interval data. *Credit: author*

WHAT PROFILING LOADS MAY REVEAL

Power use varies with time. Graphing such variables creates a two-dimensional "load profile" (see Figure C-1) with power on one axis and time on the other. Doing so may show:

- Ways to cut electric bills: by finding billing errors, or simulating another electric rate that may be cheaper.

- Identifying loads needing tighter control: seeing bumps or spikes may reveal equipment (e.g., outdoor lighting) whose operation should be better controlled.

- Securing lower retail power pricing: flattening load profiles may help do so.

- Evaluating participation in demand response programs that pay for requested load cuts: learning how loads vary over time may provide direction regarding how to control them when called upon to do so.

- Verifying time/duration of energy savings: when estimating savings from an efficiency upgrade, being able to see if the expected load reductions occurred at the expected time may help in commissioning equipment and/or verifying savings.

TYPES OF PROBLEMS FOUND/FIXED

Charting interval data has helped find and/or correct all of the following:
- Errors or failures of EMS programming
- unnecessary overnight chiller/heater operation
- failure of night setback controls
- poor training of plant personnel
- need to add controls on incremental A/C units
- outside air dampers stuck in the open position
- meter failures/dropouts (resulting in estimated bills, disputed savings, poor M&V)
- excessive interior night security lighting
- issues with on-site generation plant oversizing or dispatching
- option for splitting an electric service for lower total cost.

Applying this technique at your facility could find other ways to improve your operations and cut costs.

WHAT IS INTERVAL DATA?

Utility bills for C/I customers typically show only monthly kWh use (see upper half of Figure C-2) and peak monthly demand using electromechanical meters that must be manually read each month. Such crude data, received weeks after the fact, has limited value when trying to control a facility's energy costs.

Interval data, however, is usage information from interval (also called "smart") meters in much shorter increments of consumption, such as 15 minutes or an hour. Analyzing it (e.g., via charting or statistical methods) may reveal details in facility operations and behavior that may otherwise remain unseen. Utilities that bill for power under time-of-use rates may use interval meters to measure usage during on- and off-peak periods.

In the lower graph of Figure C-2, we see quarter-hourly interval data for one month at a commercial office building. What

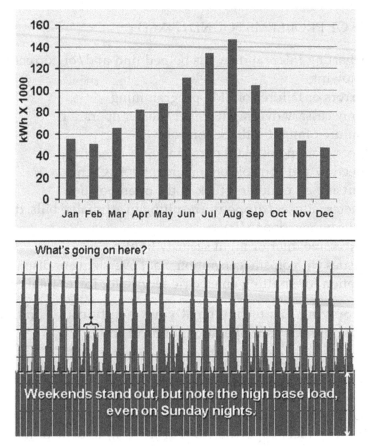

Figure C-2. Upper chart is typical *monthly* kWh consumption for a commercial building. Lower chart is *quarter- hourly* usage for 28 days in one month. The cyclical dips show lower weekend usage. *Credit: author*

do you think is happening below the arrow and bracket? If you said "weekend usage," you'd be right. The weekend/weekday pattern should be obvious: 5 days at high load followed by 2 at lower load, repeated across the one-month chart. Translating visual information into such operational data is key to quantifying and controlling loads and securing potential savings.

WHAT INTERVAL DATA IS NOT

It is not instantaneous demand. Instead, it is typically the number of kWh consumed during a defined time interval, e.g., 15 minutes. By dividing the kWh in an interval by its duration, we derive the average demand during that interval.

VISUALIZING INTERVAL DATA

Imagine stretching out one day of interval data in our example month for a closer look. In Figure C-3, we see the facility's one-day load in 15-minute intervals, with a cyclic load appearing atop the typical mound shape seen in commercial facility profiles.

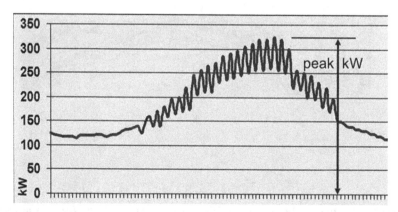

Figure C-3. Quarter-hourly load profile for a weekday in which a constant-speed electric chiller cycled to meet a commercial building's cooling load. *Credit: author*

What do you think could be causing that sawtooth shape? In this case, it's the cycling of a small (~100 ton) constant speed electric chiller. Understanding that characteristic involves integrating knowledge of a facility's systems with the charted data. Relating such charts to actual operations is much easier when knowledge of major electric loads and operating schedules is available.

An interval meter providing quarter-hourly readings will generate nearly 3,000 data points per month. Figure C-4 is a two-dimensional representation of those points, showing daily load profiles for a month overlaid on each other. The date sequence of the profiles is not visible and some profiles obscure others. Two dimensional charting of more than a few days at a time on the same chart is not an ideal way to view such data. Fortunately, several other ways are available to view many days or months of interval data on the same chart.

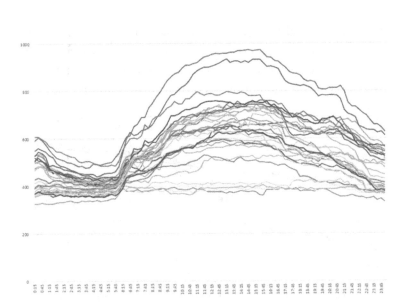

Figure C-4. Quarter-hourly daily load profiles of a college for one month. With the exception of the profiles for the two peak days, other profiles may be difficult to discern or examine when charted simultaneously in two dimensions. *Credit: author*

Energy Charting and Metrics (ECAM) is free software (developed under a U.S. DOE grant) for working with interval data. It's an Excel™ add-in that, once installed, is available for use anytime with your version of Excel™. Download ECAM and its manual at http://buildingretuning.pnnl.gov/ecam.stm. It portrays interval data in several formats.

Other programs, such as EnergyLens™, configure interval data differently. Most interval meter manufacturers offer PC-based software that works with their equipment. Several power suppliers (e.g., Constellation) offer programs (e.g., VirtuWatt) with more powerful capability (e.g., integrating hourly market pricing with hourly interval data) for sale or use by their customers.

In 1998, I developed a process using Excel™ to view a load profile in three dimensions (3D). Figure C-5 charts the same data seen in C-4 in 3D instead of 2D. I found that 3D visualization helped me analyze trends and anomalies more quickly than with 2D charts or statistical methods (e.g., scatter plots). Examples that

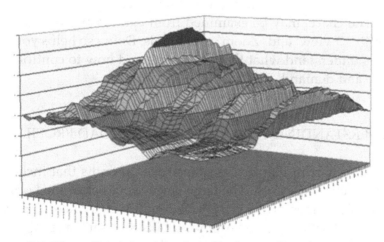

Figure C-5. Three-dimensional load profile for a college in May. Z-axis is quarter-hourly kW, foreground x-axis is days of the month, and depth (y-axis) is the 24 hours of the day. Gaps indicate weekends and/or holidays. The peak occurred when many uncontrolled incremental A/C units ran simultaneously. *Credit: author*

follow below utilize that process to portray operational issues needing attention. Several other load analysis programs now portray interval data in 3D, including ECAM.

Using templates available through AEE's online course in load profiling, the user replaces (using Copy and Paste commands) the dummy data in the Excel™ file with his own interval data. For details on the course, go to www.aeeprograms.com/Realtime/LPonline/

When that data is copied over dummy data in the templates, the new charts appear. The Z-axis is kW (or interval kWh, if not converted to kW). The foreground X-axis is days of the month (as either day numbers or actual dates, depending on how the chart is configured) or another chosen time period. The Y-axis is the time depth of the chart (as either the 24 hour numbers or actual times). Bumps and spikes (when demand is high) stand out among all the other data points.

Touching a cursor to the top edge of a 3D profile creates a tag showing the time, date, day of the week, and load at that point. That information may then be correlated with operational and weather data to assess the causes of anomalies.

The charts may be examined from various angles using Excel's™ '3D View' and 'Zoom' commands. Doing so helps you and others understand what they are seeing and how to control both usage and demand.

UNDERSTANDING WHAT WE SEE IN 3D LOAD PROFILES

The chart in Figure C-6 spans seven months that include a summer cooling season. The large empty spot below the horizontal double-headed arrow is a 2-month utility meter failure, during which the facility received estimated bills without realizing it.

Note the very tall bumps inside the oval due to the operation of an electric kiln (essentially an electric furnace). While cleared to run late at night (see the light gray vertical rectangle and spike), building staff also ran it once during midday (darker ver-

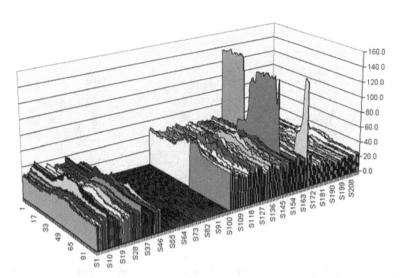

Figure C-6. 3D load profiles of a month of quarter-hourly data for a college building. Note the two-month data gap during which the utility meter had failed, and the very high plateaus due to use of an electric kiln that more than doubled the building's typical peak. *Credit: author*

tical surface) without authorization. Doing so more than doubled the monthly peak demand.

Note also that, on almost all other days, the night time demand is nearly the same as during the peak of the day. That lack of variation resulted from the temporary disabling of many EMS points during a repair, and the subsequent failure to restore their proper operation. Instead of following the daily EMS shutdown program, most lighting and HVAC systems were allowed to run all night for months.

Figure C-7 is a full year load profile for the campus of a mid-sized northeastern college. The X-axis in the foreground is the 365 days of a year. The two highest peaks (at each end of the arrow) are at the beginning (early May) and ending (late September) of the cooling season. The college partially closes in summer, so its cooling load (and kW) drops during that period.

Unfortunately, due to a lack of controls on its many incremental A/C units, it was unable to manage its load before and

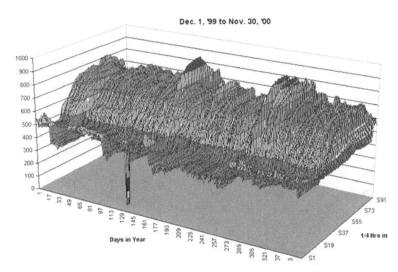

Figure C-7. In this full-year profile of a university campus, the quarter-hourly data shows the two annual peaks, just before and just after the summer (when much of the facility was closed). Operating its on-site generation for a few hours on each of those days would have reduced monthly and annual peak load, cutting monthly kW demand and other charges. *Credit: author*

after that time. If it had run about 500 kW of its emergency generators only 4 days during the year, it could have trimmed off its highest peaks, cutting its peak demand for those two months and significantly improving its annual load factor.

In Figure C-8, we see a one month (November 2012) profile for the "public lighting and power" (PLP) account in a commercial office building. It covers only the lobby and corridors on the first floor and is paid by the landlord. The spikes are due to the electric resistance heating coils in the lobby air handler. In order to warm up the lobby prior to the arrival of tenants, all of the coils come on for about an hour at the same time, creating a high peak demand charge. Instead of bringing all coils on at the same time for a short period, most of that peak could have been avoided by starting only one or two of them much earlier and running them longer. About the same kWh would be consumed in that longer warm-up cycle, and (under a TOU rate) a portion of that consumption would have occurred at the lower TOU rate. All that was needed to make that change was re-programming of the EMS and minor re-wiring of relays controlling the heating coils.

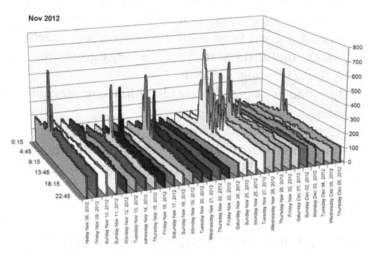

Figure C-8. In this month of quarter-hourly data, the spikes are due to brief operation of high-wattage electric heating coils for the morning warm-up cycle in the public area of a commercial building in November. By instead starting the heating process earlier in the morning using only a few (instead of all) coils, the same internal temperature could have been attained while cutting the monthly peak demand charge. *Credit: author*

Additional examples and control techniques are covered in the AEE online load profiling course.

DATA VS. INFORMATION

Utility-owned interval meters are being installed in many areas. As of 2016, over 40% of electric meters in the U.S. had interval capability. Where not yet installed, many C&I customers have installed their own. Such investment may be worthless, however, unless the data is collected, examined, and summarized on a routine basis. Without an automated system to handle such data, facility personnel may be quickly overwhelmed by it. Without good analytical capability, there may be too much data, and very little real information (e.g., trends, quantified problems). Reviewing and analyzing the data should be a task handled by a staff person with knowledge of the facility's mechanical and electrical systems, or should be done by a professional service.

Download a good (and free) manual of best practices for utilizing smart meter data at: http://energy.gov/sites/prod/files/2015/04/f21/mbpg2015.pdf.

Do the same at https://www.idc-online.com/technical_references/pdfs/
electrical_engineering/Guidance_Electrical_Metering.pdf for a free document addressing how advanced meters are being deployed at federal facilities (which include most types of buildings also found in private facilities).

DATA ACQUISITION SYSTEMS (DAS) AND SERVICES

A data acquisition system (DAS) receives, stores, and manages interval meter (IM) data. It may be an existing EMS/BMS, a separate standalone system from the meter vendor, or a Web-based service managed by a Meter Data Service Provider (MDSP). It should be able to handle more meters than are connected when initially installed and be expandable at the software level to do so. If an existing EMS/BMS is unable to handle and manipulate data, obtain meter reading software that will work with your existing computer equipment. Most meter manufacturers offer it with their meters, or as an extra product.

If the task is too much of a burden for existing personnel, consider using a MDSP. Cost for such services vary depending on whether the meter is being purchased (or simply rented), how data is communicated, and how often a customer wishes to access it. Under a typical arrangement, a MDSP installs its meters for a fee and then charges a flat rate to read and report the data, in dollars per meter per month. Rates vary widely, so it pays to compare via a RFP. A list of MDSPs certified in New York State (some serve customers in many states) may be found at: http://

www.dps.ny.gov/MSP_MDSP_Eligible.html. Your utility or PUC may have its own list for local MDSPs (some call them "Meter Data Management Agents").

DAILY LOGS OFFER INSIGHTS

Looking at past interval data helps us perceive patterns that may reveal their causation. Spotting a trend in numbers or charts is, however, only the first step in finding cost-cutting opportunities. Linking that trend to a known phenomenon (e.g., operating hours, outdoor temperature), or dismissing it as unrelated, requires ongoing knowledge of how energy is used at a facility. When chillers are scheduled to start, for example, may provide a clue regarding why power usage seen on a chart suddenly jumps. Schedules of work/occupancy/class hours, system failures or repairs, etc. may all be useful to understand anomalies.

Having such data available through daily logs and/or EMS schedules will help identify patterns that link interval data to operations and events. While manually written records (such as chiller logs) are better than nothing, reviewing such information in paper form can be very time consuming and may introduce confusion on its own (e.g., poor handwriting, missing pages, etc.). Such logs should instead be entered and maintained in a standardized digital form (e.g., .csv, .xls) to simplify extraction of data and trends.

USING LOCAL CLIMATOLOGICAL DATA TO FIND TRENDS

To correlate interval data with weather variations (e.g., temperature, humidity), digital information on local climate is essential. While many EMS/BMS include an on-site weather station to collect such data, it is also generally available (in PDF and digital forms) from public and private sources.

The National Oceanic and Atmospheric Administration (www.ncdc.noaa.gov) offers a variety of free on-line publications of its data, available for thousands of U.S. weather stations. A good starting document is the "Local Climatological Data" report. It includes average daily (and every 3 hours) dry-bulb and wet-bulb temperatures, daily degree-days, and a continuously updated 50-year comparison for that month. For a list of available data documents (some of which include hourly readings), go to https://www.ncdc.noaa.gov/data-access/quick-links#loc-clim

A FEW LESSONS LEARNED

Meters are great for collecting data, but erroneous information may be worse than no information. Before starting to use data from customer-owned meters, be sure to carefully commission the whole system (including the DAS) to find any crossed wires, software entry errors, wrong multipliers, etc.

Even utilities have failed to properly calibrate meter multipliers during their initial installation of interval meters, sometimes resulting in billing overcharges that can take months to straighten out (it took a year to fix the bills for one of my clients). A quick way to check readings is to compare them to the same months in prior years. If for a newly acquired building, compare them to energy intensity (e.g., kWh/SF/yr) data for typical buildings of the same type in your area. Find such data in the federal Commercial Building Energy Consumption Survey (CBECS) at http://www.eia.gov/consumption/commercial/

Whenever possible, monitor more than just the main utility interval meter. Isolate major sub-loads (e.g., chillers, common area lighting) on sub-meters to allow disaggregation of total load into its major components.

HOW TO PAY FOR METERS/SOFTWARE

Customer-owned interval metering may be a hard sell to superiors because there's no way to predict a payback period or annual savings from them. To cover their cost, consider these options:

- Your state energy agency may offer incentives (to either you or your utility) for installing interval meters, especially if they are to be used to participate in demand response programs.

- Some facility managers have "folded in" metering costs as part of the M&V of energy equipment upgrades that may be needed for a performance contract.

- Utilities and power suppliers may pay end users for reduced demand resulting from real-time load control fostered by such metering capabilities. The revenue from such efforts may be allocated to covering part or all of the cost of metering that make it possible.

- Find a useful document on how Uncle Sam handles the meter funding issue for his own facilities at: http://gaia.lbl.gov/federal-espc/working-groups/advanced-metering/

D: Load Factor as a Shortcut Analytical Tool

A quick way to assess the potential impact of peak demand on electric bills is to calculate a facility's load factor (LF). Find that number by dividing the average demand during an interval by the peak demand seen in a given time period. Average demand during an interval is merely kWh consumption in a defined period divided by its duration in hours.

The lower the LF, the more likely peak demand may impact your cost of electricity. In some part of the U.S., demand charges are responsible for 30% to as high as 70% of annual electric cost. If, however, demand charges total less than 20% (on an annual basis), LF may not be a significant issue.

LF is also an easy way to describe the general shape of a load profile: a low number (e.g., .3) means the peak kW is much higher than the average kW, so the curve will look pointy, probably near its middle. A high LF (e.g., .8) would indicate that the average kW and peak kW were nearly the same, so the curve would be relatively flat.

USING AN ELECTRIC BILL TO CALCULATE MONTHLY LF

Comparing LFs for different buildings is a quick way to assess which may benefit from tighter control of peak demand. Load factors are handy when comparing buildings of similar type (e.g., office building, school) but may be misleading if compared to unusual types of loads (e.g., street lighting) whose peaks may occur at very different times.

In Figure D-1, we see a monthly bill for a large (5 MW) customer. Its kWh is divided between off-peak (seen as LOPK, meaning "low-tension off-peak") kWh and on-peak (shown here are

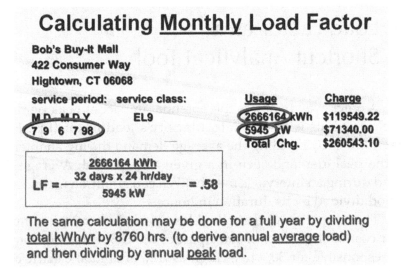

Calculating <u>Monthly</u> Load Factor

Bob's Buy-It Mall
422 Consumer Way
Hightown, CT 06068

service period:	service class:	<u>Usage</u>	<u>Charge</u>
M D M D Y	EL9	2666164 kWh	$119549.22
7 9 6 7 98		5945 kW	$71340.00
		Total Chg.	$260543.10

$$LF = \frac{\dfrac{2666164 \text{ kWh}}{32 \text{ days} \times 24 \text{ hr/day}}}{5945 \text{ kW}} = .58$$

The same calculation may be done for a full year by dividing <u>total kWh/yr</u> by 8760 hrs. (to derive annual <u>average</u> load) and then dividing by annual <u>peak</u> load.

Figure D-1. Monthly load factor (LF) may be calculated using data from a typical C&I electric bill. *Credit: author* as LOOP, meaning "low-tension on-peak"). The peak demand is shown at the level the power is delivered (i.e., the "Transmission" or "Primary" level) to be 5945 kW during the month.

That month covers June 7 to July 9, 1998, for a total of 32 days. Let's assume that the meter was read at the same time on both of those days so that this bill covers exactly 32 days. As shown in the box in Figure D-1, we sum the total kWh, and divide it by the total hours in the month to find average kW. We divide the result by the peak demand to determine the LF for that month (in this case, .58).

CAVEAT WHEN USING LF

One of the first realizations created by viewing interval data across a 24-hour period may be a surprisingly high off-peak load (e.g., due to lights and fans left running 24/7). If a facility does not tightly control its off-peak power use, its LF may be relatively high due to that wasteful operation: doing so raises average demand, and thus LF. Rather than crowing over the high LF from such waste, a facility manager should first correct the overnight operating problem and then re-assess his LF.

LF VS. POWER PRICING

Organized wholesale electric markets (except Texas) charge for generating capacity that must be maintained to ensure reliability. That cost is passed on to retail energy customers through tariffs or power contracts. The capacity charge may be based on a facility's summer peak demand, plus a reserve margin (which may range from 10% to 18%). Across a contract's term, that charge is insensitive to kWh consumption. When averaged over that term, it may add significantly to a customer's overall $/kWh price.

Load factor across a year may be impacted by a facility's summer peak demand. As a result, LF and average power supply pricing may be related: a lower LF generally means higher power supply pricing, and vice versa.

Figure D-2. As load factor increases, competitive power pricing may decrease due to the reduced capacity required to provide electricity. Most commercial buildings are in the grayed
LF range, but some industrial or
data center facilities may have
LF over .8. *Credit: author*

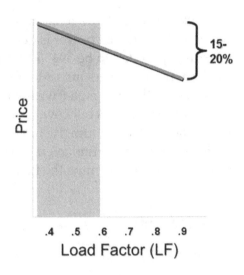

As may be seen in Fig-
ure D-2, the price difference could be significant
(though it varies among
markets and as capacity
is added or retired from
a grid). LF for most commercial and institutional buildings are
in the shaded range. If you can improve your LF by minimizing
summer peak demand, doing so may net a better price when you
seek bids for market-based power.

Where demand charges are also built into delivery rates,
a low LF may also contribute to utility monthly peak demand
charges. Reducing peak demand (and thus raising LF) may then
also reduce your average cost for power delivery.

Some utility electric rates are indirectly tied to LF. Where a
charge is based on the ratio between kW and kWh for a month,
the utility is using the inverse of LF as basis for that charge. The
higher that ratio due, for example, to a demand spike, the higher
the rate for that month

E: An Easy Way to Simplify
Your Tariff: The $/W-yr Shortcut

To accurately calculate dollar savings from upgrading or
changing their energy-consuming systems (e.g., lighting), customers need to clearly understand their rates. For simplicity,
many use the average $/kWh price derived by dividing total

annual electric bills by total annual kWh consumption. Doing so may yield seriously erroneous results.

Just as a tariff's text may be simplified, so may its math. A shortcut that I found very helpful involves calculating the annual cost of a single watt under three (or more) typical operating schedules. Annual savings from the upgrade may then be calculated by multiplying the value of a saved watt under a given schedule by the expected wattage reduction from the upgrade.

Imagine, for example, an exit sign (or toilet exhaust fan, or stairwell light) running 24/7 all year long. Let's assume the utility electric rate is \$.07/kWh on-peak 8 AM to 6 PM weekdays, \$.04/kWh off-peak all other times (both including the various \$/kWh small fees seen on a typical bill), and demand is charged at \$10/kW per month. Annual cost of consuming one watt is then:

(\$.07/kWh x {52 wk/yr x 10 hr/day x 5 days/wk}) + (\$.04/kWh x [8760 hr/yr—{52 wk/yr x 10 hr/day x 5 days/wk}]) + (\$10/kW-month x 12 months); divide the sum by 1000 to convert kW into watts, yielding (\$182 + \$246.40 + \$120)/1000 = \$.5484/W-yr.

If switching from a 50-watt incandescent exit sign to a 7-watt LED exit sign then avoids 43 watts, the annual savings would be 43 W x \$.5484/W-yr = \$23.58. If the unit's installed cost is \$47, we get a payback period of about \$47/\$23.58 = 2 years. When the *average* \$/kWh price was instead used, a longer payback period resulted.

Perform the same analysis for a watt saved for these two other schedules to see how much annual cost may differ based on when the watt is used:

- 9-5 weekdays all on-pk (e.g., general lighting, constant speed air handler fans)
- 4,000 hr/yr off-pk only (e.g., occupancy sensors, outdoor lighting)

Develop your own schedules that better fit your facility's operations.

Once done for each typical schedule, this process avoids re-doing a laborious tariff calculation every time a new energy upgrade is being considered. While not perfect, the result may be significantly more accurate than using average $/kWh.

F: Options for Professional Assistance With Competitive Procurement

To help choose a supplier, a customer may use a broker, consultant, and/or online auction service. For simplicity, we call them "specialists."

A *broker* is an individual or company having arrangements with multiple (but not necessarily all) power suppliers to act as their sales agent. Most states require that brokers be licensed (exceptions include New York and Michigan) and list them at their PUC web sites. Their service tends to be relatively narrow: find the supplier offering the lowest fixed price for a customer. The broker will receive a commission from the winning bidder that is proportional to the quantity of kilowatt-hours (kWh) bought by the customer. Measured in mills (i.e., tenths of a cent) per kWh, that fee is added to the price the supplier would have charged the customer if no broker was involved. No money passes directly between broker and customer, allowing brokers to claim their service is "free." A high commission may, however, consume a significant portion of the savings otherwise available to a customer.

Unless the customer asks to know the broker's commission, there is no requirement that it be revealed. The broker is typically paid month-by-month, based on the customer's monthly kWh consumption. Some suppliers pay their brokers annually in advance, with a true-up at the end of each year. A broker therefore takes some risk that the customer will stay in business (and pay its power bills) during the term of the contract.

Broker fees vary widely, depending on the size and complexity of the customer, with smaller customers being charged a higher mill/kWh rate. A mid-size commercial or industrial customer (e.g., 500 kW peak demand using ~2 million kWh/yr), might incur a broker's fee of 2 to 3 mill/kWh, totaling $4,000 to $6,000 (i.e., $333 to $500 per month) spread across a one-year contract term. Such fees may add 1% to 2% to a typical total electric bill (including utility delivery charges). Larger customers (e.g., 5,000 kW peak demand using 20 million kWh/yr) should seek a lower mill/kWh fee.

Broker fees are not regulated. While some suppliers limit fees to about 5 mills/kWh, at least one allows up to 10 mills/kWh (i.e., $.01/kWh) which may result in the customer seeing little or no savings relative to his utility's supply pricing. If asked, some suppliers may tell a customer the broker's fee, though some have non-disclosure agreements with their brokers barring them from doing so. A customer may ask a broker to show his contract with a supplier wherein the commission is shown. A refusal to do so may be a reason to choose a different broker.

A customer being approached by someone who says "I can get you cheaper power at no cost to you" should first ask to see that person's brokering license, if in a state that requires licensing. That name should then be checked at the PUC web site to confirm the license is currently valid. A broker's license may be pulled for violations of commercial behavior, but such cases are very rare. Multi-level marketing to sell retail power is common in many states. The person claiming to be a broker may, in reality, be a subordinate to a licensed person or firm, adding yet another fee to your power price.

Many brokers offer their services without a written contract with the customer. That's a bad idea for both parties: with no agreement to do otherwise, a customer may play multiple brokers against each other, and (with no written contract) a customer cannot be certain what services he will get, nor what recourse he may have if dissatisfied with the service. Unless clearly listed in a broker's written scope of work, other tasks related to the transaction (e.g., bill review, supply contract analysis) may not be included.

Potential conflicts-of-interest also need to be considered. A broker being paid by suppliers may not be the best person to critique their contracts, and may steer a customer to a supplier offering the broker a higher commission. Because the commission is proportional to a customer's kWh consumption, an incentive also exists to push a customer into a multi-year contract that ensures an extended cash flow for the broker.

To deal with such issues, some suppliers only allow one of their brokers to serve each potential customer. That too has a downside: if a customer tries to deal directly with a supplier or wishes to choose his own broker, he may instead find that his account has been "allocated" to a broker he never heard of. It is a common broker trick to accumulate account numbers of customers he never met, and then claim exclusivity to block access to those accounts by other brokers. Should that happen, a customer may undo that limitation by notifying (on letterhead) the customer service reps of major suppliers that it has not granted exclusivity to any specialist (or listing the one to whom exclusivity has been granted).

While not legal, some brokers "slam" customers by switching their accounts to their preferred supplier without the customer's knowledge. To do so, they need the account number and service address, both of which appear on each utility bill. Never give your utility account number or a copy of an electric bill to a broker (or anyone outside your firm) without a written memo (signed by the recipient) limiting how it may be used.

Some utilities allow customers to "block" their accounts, meaning that the utility will not accept a change of supplier (whether utility or non-utility) without written confirmation from the customer. If interested in that option, contact your local utility account rep.

With all those caveats, why use a broker? Some types of customers (e.g., commercial real estate) would rather do so than try to find low pricing on their own. There's a lot to be said for a specialist who brings expertise and experience to the table, but never submits a bill for it. And if the fee gets passed on to tenants in their electric bills, a landlord is essentially indifferent to it. A landlord may then fairly claim he secured a service to get the best price for his tenants at zero cost (to the landlord).

An *aggregator* gathers and bundles groups of customers to bring to a supplier. In many ways, they are essentially brokers. At least four states, including Texas, require aggregators to be licensed. Like brokers, aggregators receive commissions, though the formula may differ depending on the type of aggregator.

Aggregators serve a useful function for small customers whose kWh volumes are too low to have negotiating leverage, or may otherwise incur higher brokering fees. Some existing customer organizations (e.g., BOMA chapters, chambers of commerce, hospital associations, industrial alliances) act as aggregators for their members. They receive a commission based either on the total annual customer kWh volume or some other formula (e.g., $X0,000 per year) to bring members to a supplier. In some cases, the aggregation group may use a broker to help it find the best price. The winning supplier may pay separate commissions to the broker and the organization, or the broker may split the fee it receives with the organization.

Towns and cities (e.g., Chicago) in some states have, through referenda or legislation, secured the right to act as aggregators for their residents, including homeowners and businesses, whether or not those customers desire it. Called "municipal (or community) aggregation," this method allows customers who prefer to remain with the utility as their supplier, or wish to seek suppliers on their own, to opt-out through a process. Otherwise, their procurement is managed by the aggregator. A municipal aggregation allowed a small city to collect a 1 mill/kWh fee for all power it secured for residents and businesses that didn't opt out. That resulted in a commission to the municipality of several million dollars a year—just the trick to plug a hole in its annual budget.

Like brokers, *consultants* help customers find competitive power pricing. Anyone can set himself up as a consultant: no state requires them to be licensed (recall the adage that "a consultant is any guy from out-of-town with a briefcase"). Unlike brokers, they are paid directly by customers, typically under fixed fees for a defined scope of work. That fee may be proportional to the annual kWh volume, number of accounts, and other factors. Consulting fees vary widely but tend to mirror those of brokers.

Consultants prefer to serve large customers (those using at least several million kWh/yr) and to collect—usually in a lump sum—a commission when a power deal is done, rather than receiving small payments each month. Unless a consultant also has brokering arrangements with suppliers, it has no incentive to limit the competition to those offering higher brokering fees, or to push for longer terms to secure future cash flow for the consultant. Unlike a broker's commission, a consultant's commission should not be proportional to the term of a power contract unless a multi-year deal entails much more work or complexity. If paying a consultant under a fixed fee arrangement, contractually bar it from also collecting a brokering or online auction commission.

To get the most out of a consultant, task it to look broadly at ways to cut energy pricing, from both competition and other means, via a defined scope of work. Some of the tasks listed below may also be appropriate to a brokering contract. Some may each take only a few minutes, while others could take much longer.

- help assemble accurate account data (i.e., numbers, service addresses, access passwords)
- check suitability of existing delivery tariff(s) and taxes
- benchmark annual usage/demand against similar local facilities
- for large accounts, analyze hourly load profile(s) if such data is available
- review year-to-year consistency of account usage
- assist with credit issues or financial standing
- explain pricing structures and risks
- offer options for price discovery methods (e.g., online auction, sealed bids)

- critique bidder contracts prior to pricing
- forecast utility and bidder pricing
- watch market for pricing opportunities
- model total costs of proposals
- ensure simple signing procedure
- aid in securing non-price benefits.

Before engaging a consultant, consider these issues.

- What are you able to spend on a consultant? His fee may exceed what you have in your budget for consultants.

- Ask how his fee will be calculated (e.g., mills/kWh) and in the consultant's contract be clear that it cannot also include a separate brokering commission that increases the winning supplier's bid.

- Look for verifiable procurement experience, not company size or claimed "connections." Review its knowledge of your utility's tariffs, load analysis, negotiation of power contracts, etc.

Find a list of certified energy procurement consultants (who have taken a 3-day class and passed a 3-hour test) at www. aeecenter.org/custom/cpdirectory/index.cfm At its first page, enter your state or province, nation, and (in the certification choice field) choose "CEP—Certified Energy Procurement Professional." Other fields may be left blank.

Online reverse auctioneers work a lot like eBay (except that the lowest bidder is the winner) and are a popular price discovery method for many types of products. Online energy auctioneers are essentially brokers. Each is active nationally or regionally and may hold multiple state licenses. Like a broker or consultant, an online auctioneer handles the procurement process. Its main feature is a live on-screen auction visible to all competitors and the customer. Within a defined period (or number of bids) usually lasting less than an hour, such events offer both excitement and clear evidence of competition.

Brokers and consultants may use an online auctioneer as part of their services. Like a broker, the auctioneer has arrangements with multiple (but not necessarily all) suppliers and collects a brokering fee from the winner. It may share that fee with the broker or consultant that brought the customer to its service. Some consultants have been known to use online auctions as a backdoor way to collect a hidden additional commission without needing a broker's license.

Some auctioneers claim to perform contract review and other procurement services. Such analyses may not be as rigorous as a consultant because, like brokers, an auctioneer is compensated by the winning supplier based on a $/kWh fee. Due to past disputes, some power suppliers will not work with some auctioneers, thus limiting the customer's choice of suppliers and price competition.

The customer's load and contract terms are posted at a secure web site. After multiple rounds, the lowest bidder is the winner. Worldenergy.com (owned by EnerNOC, a major demand response provider) handles most online energy auctions. Others include www.procurexinc.com, www. coexprise.com, and www. energymarketexchange.com. Some brokers and consultants use their own online auction software.

CHOOSING A SPECIALIST

Like any other service, power procurement assistance may be subjected to competitive bidding. That option works best when customer (or aggregation) kWh volume is large. Consultants compete by lowering their proposed fixed fee, while brokers may quote a lower mill/kWh rate. When competing with other consultants, I have been stunned at some quoted fees. In one case, a large consulting firm (with an apparently very high overhead) wanted to charge almost 5 times what other consultants were willing to accept for the same work.

A good specialist will, before asking a customer to sign his contract, help the client understand how power supply pricing is structured and the available procurement options. Those that are instead stingy with their knowledge may be revealing a lack of it.

WHEN NOT TO USE A SPECIALIST

There are situations where paying for a specialist may not be worth the money. Specialists often do well with high kWh volume accounts, but not with many small accounts, especially if they are spread behind multiple utilities. In such cases, using a specialist may be overkill.

"Small" would include accounts where an annual total bill (supply + delivery) is less than $20,000/yr. Such accounts may have annual peak loads under 50 kW. Instead of spending resources on a formal RFP and bidding process, it may be more cost-effective to simply join an existing aggregation group, or seek informal bids from several of the larger suppliers.

Unless joining an aggregation group, professional due diligence requires that a customer review financials from potential suppliers regardless of customer size. Suppliers that refuse to provide basic information (such as that discussed in the RFI section of Chapter 13) may have something to hide and should be excluded from consideration.

Definitely use a specialist for multiple accounts each spending at least $100,000/yr (i.e., roughly 300 kW peak per account), or any combination spending over $1,000,000/yr. For a big spend, (e.g., $5,000,000/yr or more), negotiate a low fixed commission or fee (e.g., .3 mill/kWh or less).

DOING IT ON YOUR OWN

Training courses and a professional certification in energy procurement are available through the Association of Energy

Engineers (www.aeecenter.org). Live 3-day classes covering both wholesale and retail power and natural gas procurement are given at AEE national events. Details on its online retail power procurement course are available at www.aeeprograms.com/realtime/PowerPurchasing.

G: Buying On-site Energy from a Competitive Supplier

INTRODUCTION

Over 9% of all electricity consumed in the U.S. is generated through on-site power plants. Two types of power systems dominate this field:

- cogeneration, also called Combined Heat and Power (CHP), provides both electric and heat energy (some systems that also provide cooling are called "trigeneration")

- on-site renewable power sources, such as solar photovoltaic (PV) panels or wind turbines, provide only electricity.

Most existing on-site kW capacity in the U.S. is in CHP systems at large industrial and institutional facilities, though some universities, industrial parks, and even housing complexes have also adopted this strategy. A CHP system supplies part or most of a facility's energy (i.e., power and heat) needs. The balance is covered by the utility (or a deregulated power supplier) and the facility's own boilers. A customer buys the power and the generator's waste heat at a discount relative to what it would otherwise have paid. From the customer's standpoint, that's like getting a cost-free rate cut.

Many C&I customers are installing or hosting large solar PV systems that provide on-site power, sending excess PV power into the utility grid. Depending on limits in the local utility tariff, a PV system may—over a year's time—provide more kWh than consumed by a hosting facility, effectively zeroing out its kWh consumption. The facility's electric bills may then consist primarily of its fixed and demand-related charges.

The high capital cost of such systems has led many to host (rather than own) private power vendors who finance and install their own equipment at customer sites. An agreement between a hosting customer and an on-site power vendor is called a Power Purchasing Agreement (PPA), though that same term is also be used to describe other forms of power contracts.

Contracts for buying such on-site third-party energy may be complicated, with terms lasting 15 to 25 years. All eventualities (e.g., bankruptcies, asset transfers, changes to energy use) must be clearly spelled out, yielding PPAs that run 20 or more pages. When considering such an arrangement, following are some of the concerns that need to be addressed.

CALCULATING SAVINGS FROM ON-SITE GENERATION

One of the financial benefits of on-site CHP generation is the avoidance of utility delivery charges proportional to kWh usage, which may be ~50% (or more) of an electric bill. Even if the cost to generate on-site power is higher than a utility's supply rate, net savings may result from avoiding delivery charges that would otherwise have been paid if that power came instead from a utility. An on-site CHP system also produces waste heat usable to warm a facility's space, domestic hot water, and/or process loads (e.g., swimming pool, laundry, industrial heat load). Buying that thermal energy at a cost lower than the incremental cost from using a facility's boilers may also provide significant savings. Waste heat may also be used to produce chilled water (e.g., from an absorption chiller), thus reducing kWh consumption of electric chillers.

Careful attention is needed, however, to ensure such dollar

(as versus Btu or kWh) savings actually occur.

Issues have arisen with CHP systems whose power is sold at a fixed $/kWh price discounted off the customer's prior average electric rate. That price would thus include whatever the customer had paid for peak demand charges to the utility. It may implicitly assume the CHP system will always be providing its full kW capacity during every month covered by the contract.

If the system instead shuts down while demand is high, the utility may levy a demand charge for the kW load that was not provided by the CHP system. Some PPAs allow an on-site power vendor to shut down the CHP system when the daily wholesale price of natural gas (the primary CHP fuel) spikes. The customer could, for each month such a failure occurs, end up paying twice for his peak demand: once to the CHP vendor (as part of its fixed contract price) and again to the utility. If a failure or shutdown occurs at the time the local ISO peaks, the customer could end up paying a much higher annual capacity charge every month for a full year after the CHP failure. If the replacement kWh from the utility is on a TOU or RTP rate, an extended on-site outage could also lead to higher-than-average supply rates for that electricity since it must be made up by the utility at its tariff rate.

At least two ways exist to approach this problem. Avoid it by metering and paying separately for the kWh and the kW (based on whether the utility demand charge was indeed avoided each month). During contract negotiations, customers may find that the CHP vendor refuses to allow such a transfer of financial risk, causing the deal to fall through. Alternatively, a utility may offer (or be willing to create) a tariff that allows for a few CHP failures each year without charging for the resulting extra peak demand, in exchange for the customer paying a fixed "contract demand" charge whether or not the CHP system always works properly. That charge may, however, consume some of the dollar savings previously claimed by the CHP vendor.

BTU METERING IS PARAMOUNT

When buying energy from an on-site vendor, proper metering is key to ensuring accurate billing. For small CHP systems, Btu metering of heat provided to the customer is accomplished by measuring the flow rate (in gallons-per-minute, gpm) in the hot water loop from the generator and multiplying it by the temperature difference between the supply and return pipes. For larger systems that instead produce steam, steam meters would be appropriate. In either case, annual verification of accuracy (especially for steam meters) is essential. The cost for that effort should be built into the CHP vendor's rate rather than being charged separately.

Problems may arise, however, regarding the location of such metering. In one case, the vendor metered at the inlet and outlet of the generator, assuming that all heat taken from the loop was used by the host facility. Unused waste heat was, however, being discarded out of a cooling tower (needed to ensure proper operation of the generator) on the same loop. The end result was that the total energy bill (electricity plus heat) was higher than if the CHP system had never been installed. A nasty lawsuit resulted, ending in removal of the system and cancelling of the contract.

BE CAREFUL PROJECTING FUTURE ENERGY PRICING

Under an assumption that utility rates will rise over time, an escalator (e.g., 2%) may be annually added to the vendor's energy price. While history has indeed shown a relatively continuous increase in power pricing by many utilities, assuming a constant price escalator on top of a rate picked at some point in time may not be the best way to project future rates.

During the recent recession (roughly 2008-2014), a sustained drop in grid demand—coupled with a large and sustained drop in natural gas pricing due to new drilling technologies—yielded a significant drop in the value of the supply component of utility electric rates.

If a PPA had locked in a 10% discount and a 3% annual escalator at the height of that power pricing (applied to the total price
paid for power), the customer would, over the next 7 years, have
paid about 50% more than he would have if the system had not
been installed (assuming that the utility's *delivery-only* price had
indeed risen 3% per year).

A better approach could peg the annual CHP price to the
rate the customer would have paid without the system, based on
the utility's published rates, trued up at the end of each year. If
the utility rate continued to rise (as projected by the vendor), so
would the CHP rate. But if the utility rate went down, that price
risk would be borne by the vendor, not the customer.

BE AWARE OF EXTRA LIABILITIES

To minimize the cost of natural gas (the fuel of choice for
most CHP systems), a vendor's design may include a backup fuel
(oil or propane) tank to secure a lower delivery rate from the local
gas utility. On-site fuel storage could expose the host to fines or
penalties if the tank leaks, and may require attending to inconve-
nient deliveries.

A gas-fired generator may require either high pressure gas
piping or a gas pressure booster, either of which may entail additional building code approvals and perhaps an increase in liability insurance for the facility.

If the generator is noisy (but not in violation of codes) or
causes distracting vibration, neighbors (or tenants) may complain. Costs for correcting the problem (e.g., acoustic barriers)
may then fall on the customer, not the vendor, unless stated otherwise in the PPA.

A customer may find that its facility's boiler emissions permit could be jeopardized or require expensive revision due to the
additional site emissions from the CHP unit.

At one site, a commercial real estate firm wanted to use the
space occupied by a CHP system for a well-paying (and growing)

data center tenant. Under the PPA, the customer would have to pay to re-locate the CHP system to the roof (very expensive) and for any revenue lost by the vendor while the system was disabled. That cost was too high, and the landlord lost the lucrative tenant.

SECURING EMERGENCY POWER AS PART OF THE DEAL

Merely having a generator on site does not ensure power during a utility outage. Many small-scale (i.e., less than 1,000 kW) CHP systems use generators that may not be well suited to act as blackstart generation (i.e., starting and running on their own when utility power has failed). Unless otherwise designed, both CHP and PV systems run in parallel with a utility's power output. They shut down during an outage to avoid backfeeding power into the utility distribution system at a time when workers may be repairing it. If blackstart is essential, achieving it through a PPA may involve additional generation and distribution equipment, the cost of which falls on the customer, unless blackstart is included as a service in the PPA.

AVOID LIMITS ON FUTURE OPTIONS

Many PPAs contain language that limits a customer's right to reduce electric or thermal usage, e.g., from improved energy efficiency or changing plant equipment. Blocking that ability ensures a steady revenue stream and profit for the vendor. In the case of CHP, all of the following have been barred or limited in some PPAs: PV panels, efficiency upgrades, non-electric chillers, thermal storage (which may shift the time of peak load to hours when waste heat cannot be fully utilized), and demand response. If space heating from a generator's waste heat is a big part of the vendor's revenue equation, even installing more efficient windows may be off-limits. For cost-conscious customers seeking to reduce their energy bills (and/or carbon footprint), such limitations—for 15 to 20 years—may not be tolerable.

THE SPECIAL CASE OF PV POWER

Some customers that hosted large PV systems have watched their billed kWh usage drop noticeably, but not so their total electric bills. Due to a mismatch between the facility's *load profile* and the PV system's *output profile*, monthly peak demand may not be reduced to the same degree as billed kWh. When the PPA's price is discounted off of a facility's average $/kWh that had the costs of peak demand folded into it, the actual value of the PV power may be less than the power it displaces, resulting in a higher total bill (i.e., utility plus PV vendor). If battery power storage costs come down sufficiently, this problem may be avoided, but the cost of the batteries and their gradual loss of output over time need to be factored into the economic calculations.

**The Sun May Always Be Bright,
But It May Not Always Be Profitable**

An industrial site (5+ MW peak demand) with lots of flat roof space was approached by a PV vendor offering to install a 2 MW system at no charge to the customer. Under the deal, the customer would buy all power generated by the system, with the remainder coming from the utility. After reviewing the customer's electric bills, the vendor offered fixed-priced power at $.15/kWh, which was about 10% lower than the customer's average electric rate at the time.

The customer's energy consultant pointed out, however, that the facility's interval meter data showed that (due to its industrial operations) it peaked on most days before 10 AM, long before the PV system's output was significant. As a result, peak demand would not be lowered to the same degree that kWh taken from the utility would be displaced by the PV system. Because the $.15/kWh price was a discount off of the customer's average historical rate—which

---

included peak demand charges—it was not an appropriate way to value the kWh from the PV system.

The facility's average rate without demand charges was only about $.10/kWh. To properly reflect the actual value of his PV power at the site, the vendor was asked to drop his proposed price to 10% off that $.10 rate. He indignantly refused, even as he conceded the validity of the interval data analysis. He stated "we sell lots of PV power to others in your area at $.15! Nobody else ever brought up such issues!" He was politely shown the door. Another PV vendor later offered a discount off the $.10 price.

Had the customer instead signed the proposed 20-year deal at $.15/kWh, he would have lost (not saved) over $2 million during the term of the PPA.

H: Low Risk Interventions in Rate Proceedings

Some customers eager to enter the rate proceedings arena make mistakes resulting in both disappointment and loss of credibility. Here's an important tip: before proposing a tariff change, be sure it does not already exist in the tariff or is not already available as a service by request to the utility. Doing otherwise has left some intervenors looking like fools during a proceeding.

Before taking on ambitious actions (e.g., challenging a cost-of-service study), consider pursuing a few of the following to gain experience while making less-than-spectacular improvements to utility services and options. Many have been won by energetic customers. Some may also come in handy as "bargaining chips" to be negotiated away in exchange for something else.

- require quarterly publication of all non-bypassable utility charges (including their $/kWh value) via a press release and/or a bill insert delineating them

- create an option for "calculated" bills to be provided upon request that break out every charge, instead of bundling them as is typical in monthly bills (see Figure H-1)

- annual delineation of claimed line loss (a.k.a. unaccounted for energy [UFE]) by rate class, with a penalty on the utility for exceeding X%

- access to real-time cost-free interval data, or accelerated/low-cost installation of such metering

- provide electronic invoicing, e.g. EDI, upon request at no extra charge

- offer incentive for early payment via EFT

- if a tariff allows sub-metering of tenants, limit the percent markup that may be charged above the monthly average utility rate

- lengthen the period for recovery of utility overcharges from billing errors (many cut off at 2 years)

- penalize the utility or credit customers for utility outages (not due to extreme weather) lasting over 4 hours

- instead of a required deposit, allow use of a bank Letter-of-Credit (LoC)

- all billing "adjustments" over $X,000 to be delineated in writing, along with their calculation

- add a penalty on the utility for a meter failure lasting more than one billing cycle

- allow coincident/conjunctive demand metering for a customer's accounts located on the same site, or served by same high-tension feeder, or same substation, or same distribution transformer

- exempt customers who choose not to participate in energy efficiency programs and those operating zero energy buildings (ZEB) from non-bypassable charges supporting those programs

- reduce penalties for brief failures of on-site cogeneration (and/or provide credits for no failures)

- provide free or low-cost assessment of costs related to interconnection of on-site generation

- link fuel adjustment charges to a publicly available index (e.g., NYMEX gas, ISO day-ahead)

- reduce the late payment penalty rate (standard 1.5% per month = 18% per year, which is too high)

- if a TOU and/or RTP rate is available, require that the utility provide a free analysis of the differential cost or savings relative to the customer's existing rate using its data from the prior year

- institute virtual remote net metering that allows a customer with on-site generation (e.g., PV) to allocate excess output to its other accounts

- unbundle rates to separately show (e.g.) high- and low-tension charges so customers may easily calculate the value of switching to high-tension or transmission level service

- add an option to buy renewable power from the utility (and thus avoid an existing RPS charge)

- replace programs that charge all customers to subsidize low-income residentials with an "energy stamp" program (akin to food stamps) paid out of the budget of a social program instead of by ratepayers

- if stranded costs are still being paid due to deregulation, require that they appear as a separate charge, and that an annual accounting of such payments be made public, highlighting any additional charges that "snuck in" to the total

- create an incentive for installation of power factor correction equipment

- adjust tiered rates based on actual cost-of-service, rather than as a form of social engineering

- provide a risk-free RTP that provides hourly pricing, but (during a 12-month period) ensures no net loss compared to what a customer would have paid on standard rates

- if a state has a RPS, push for a feed-in tariff (FIT) or a solar REC program to allow customer participation in fulfilling it

- costs to comply with a RPS may be built into electric bills as non-bypassable charges; a customer who on its own (or via its supplier) buys RECs covering 100% of its power, or secures part or all of his power from an off-site renewable source, should be exempt from RPS charges. Otherwise, it's paying double for renewable power.

- request a business incentive rate for your rate class if it produces jobs and/or revenue in the community

- where deregulation exists, require that the utility offer a free analysis showing how much a customer would have paid if it had instead remained on the utility's supply rate in the prior 12 months

- raise the value of demand response incentives so that it includes avoided T&D costs, as well as those based on the hourly commodity market

- if not already existing, seek a separate rate for street, parking lot, and bridge lighting controlled by photocells; since such lighting is typically not on during a utility's or ISO's peak load, it should not contain any capacity charge

- in still-regulated states, offer a 'green' tariff that allows a customer to access off-site renewable generation from a third party

- create incentives for installation of (e.g.) energy storage, demand controls, LED lighting, etc.

- provide favorable start/end times for TOU rates

Customer Name: _____
Address: _____

ACCT # _____

FROM: 06/21/12
TO: 07/23/12

ELECTRICITY FOR   32 DAYS
SC 39 RETAIL CHOICE   RESIDENTIAL USE   0%
ISO LOAD ZONE J

RATES EFFECTIVE: 05/31/12 TO 09/30/12
SC9-GENERAL LARGE SUB-METERED

LOW TENSION ON PEAK CALCULATION
    MARKET SUPPLY   1637964.0 KWHR @ $ .000000   .00000
    MONTHLY ADJUST   1637964.0 KWHR @ $ .013645   22350.01878
    DELIVERY   1637964.0 KWHR @ $ .008200   13431.30480
    SBC / RPS   1637964.0 KWHR @ $ .004900   8026.02360
                                                   --------   ------------
    SUBTOTAL   0.026745   43807.35000
UNCOLLECTABLE DELIVERY   1637964 KWHRS @ $ 0.000087   142.50
TRANS ADJUST - DELIVERY   1637964 KWHRS @ $-0.000064   -104.83
RDM ADJUSTMENT   1637964 KWHRS @ $-0.003486   -5709.94
PSC ASSESSMENT CHG   1637964 KWHRS @ $ 0.003945   6461.77
ADJUSTMENT FACTOR   1637964   KILOWATT-HOURS @ $-0.000505   -827.17
GRT COMMODITY TAX CHARGE $ 0.00 @ 2.4066%   0.00
GRT DELIVERY TAX CHARGE $ 43769.69 @ 2.4429%   1069.25

    SUBTOTAL $ 44838.93

LOW TENSION OFF PEAK CALCULATION
    MARKET SUPPLY   1711636.0 KWHR @ $ .000000   .00000
    MONTHLY ADJUST   1711636.0 KWHR @ $ .013645   23355.27322
    DELIVERY   1711636.0 KWHR @ $ .008200   14035.41520
    SBC / RPS   1711636.0 KWHR @ $ .004900   8387.01640
                                                   --------   ------------
    SUBTOTAL   0.026745   45777.70000
UNCOLLECTABLE DELIVERY   1711636 KWHRS @ $ 0.000087   148.91
TRANS ADJUST - DELIVERY   1711636 KWHRS @ $-0.000064   -109.54
RDM ADJUSTMENT   1711636 KWHRS @ $-0.003486   -5966.76
PSC ASSESSMENT CHG   1711636 KWHRS @ $ 0.003945   6752.40
ADJUSTMENT FACTOR   1711636   KILOWATT-HOURS @ $-0.000505   -864.38
GRT COMMODITY TAX CHARGE $ 0.00 @ 2.4066%   0.00
GRT DELIVERY TAX CHARGE $ 45738.33 @ 2.4429%   1117.34

    SUBTOTAL $ 46855.67

LOW TENSION 8AM-6PM DEMAND CALCULATION

Page 1

Figure H-1. First page of a "calculated" utility bill showing how the cost of many components was determined. Two other pages showed details on other components. To better understand your electric bill, ask your utility to provide such detail upon request. *Credit: author*

- create a separate manufacturing/industrial rate for high LF customers

- standardize interconnection procedures and rates for non-utility on-site power generation

- make all "special" and "off-tariff" rates subject to public disclosure under the state's Freedom Of Information (FOI) law since they were developed and approved by a government agency (e.g. PUC).

Brainstorm other options that you and your allies might enjoy if available. After a few successes, consider more adventurous options, such as challenging a cost-of-service study that allocates too much cost to your rate class (some label this "rate discrimination"), or creation of an interruptible rate, or seeking rate designs that encourage—not discourage—on-site generation.

I: Forming and Using an Energy Buyers Group

INTRODUCTION

Energy customers are pooling their talents through the creation of buyer groups, purchasing consortia, and industrial councils. Many work together to cut their energy costs by representing their common interests before public utility commissions, utilities, courts, statehouses, and energy vendors. Many have found that doing so can literally change the rules.

Before considering creation of a new customer group, review the lists of intervenors in recent rate proceedings. They may be listed in documents (such as rate settlements) posted at PUC web sites. You may find allies already acting as intervenors, and decide to join their group instead of creating a new one.

WHY ORGANIZE?

Unless a customer's electric load is very large, it may have little or no influence over how energy is priced, regulated, and sold. While they may grumble, most customers accept whatever rate increases are levied upon them. The deregulation of phones, airlines, natural gas, and power have, however, demonstrated new methods to confront rising utility rates.

The key to success is to demonstrate that many customers want change, and will spend their time, funds, and attention to make it happen. When intervening before utility commissions, customers often have greater credibility than equipment vendors or energy suppliers, especially when a group includes public agencies, nonprofit institutions, and well-known religious, health, or educational facilities.

A SHORT CASE STUDY

A large user of electricity and natural gas in a major U.S. city sought access to an economic development electric rate for the new central plant it was constructing. After filing lengthy paperwork to secure it, an error by the power company, related to interconnection of electric service to the new facility, jeopardized obtaining the new rate. Losing access to this benefit could have cost over $6 million in the next decade.

Lengthy correspondence and meetings with the utility produced no results, so the customer retained an attorney, well versed in utility case law, to bring an action on the matter. Suddenly, the utility "discovered" a way to reconfigure the electric service that would avoid the problems it had created. The threat of legal action was dropped without the customer incurring any fees.

EFFORTS TO EXTEND THE SUCCESS

Surprised by the ease of this success, the customer investigated other ways in which legal action, or the threat of it, could help cut utility bills.

Discussions with several consultants on utility issues was not, however, very productive. Despite their often heated claims to the contrary, many were insufficiently knowledgeable (or did not fully understand) prevailing rate tariffs. Others had become so closely involved with the regulatory process that they had gradually "bought into" the often dubious arguments made by utilities. They had little incentive to alter the system because, win or lose, they collected their consulting fees, much as stock brokers collect a fee from every trade, regardless of the profit or loss involved. A different catalyst was needed to encourage action.

SEIZING A NEW OPPORTUNITY

Over the next two years, the customer's energy manager investigated ways to obtain reductions in natural gas pricing. In the late 1980s, a tariff had been created that allowed the local utility's gas lines to deliver gas from other sources, but it had been used by only three large industrials with easily controlled, process-related loads.

As a result of gas deregulation, several natural gas suppliers were offering new supply options. All tried to tie the customer to wholesale-style "take or pay" contracts, or other unwise maneuvers. Only one was willing to enter into serious negotiations that took into account the weather-sensitive nature of a large institutional facility. After six months of negotiations, a groundbreaking contract was developed.

Since the only other source of natural gas was the local utility, it was given an opportunity to compete with the supplier. This concept was (in early 1991) so foreign to the utility that it was unable to offer a coherent response, instead making veiled threats regarding supply security and its own displeasure. The only leverage left to the utility was to try to scare the customer by exaggerating the uncertainties inherent in any new venture.

Ignoring these baseless claims, the customer plunged ahead,

saving over $1,000 a day (in 1991 dollars) under its new gas contract.

EXTENDING THE OPTION

By exploiting this little-used tariff option, the customer set an example for other institutions and large gas users in the area. Soon, hospitals, universities, and even the city's government asked for its help in accessing cheaper natural gas. Within a year, over a dozen were buying their gas on the monthly spot market. A close relationship formed between the supplier and the customer.

When those customers saw other ways to improve the existing gas procurement process, it became clear that action before the state public utility commission (PUC) might be needed. Experience during the gas contract negotiations demonstrated the limitations of using in-house legal staff, most of whom were experienced in other aspects of law. Paying for outside legal help was also problematic due to budgetary restrictions. To find other spot gas users with whom the cost of a rate intervention (and its benefits) could be shared, the customer asked its gas supplier for their names, with the promise that the list would be kept confidential. Those customers were called to a meeting to discuss recent utility billing and customer service problems all had been experiencing. From this first meeting emerged a new buyers' group.

To represent itself before the PUC, the group retained the services of the firm used in the customer's electric rate case. A dues structure was created that charged based on gas usage, with those benefitting the most from changes paying the most. Bylaws were written, officers elected, and meetings scheduled. The group was then ready to develop and pursue united actions to benefit its members.

During its first year, the group was instrumental in obtaining improvements to its members' billing, shared pricing and purchasing information, and was well situated to act when the utility offered its own version of a spot gas contract. Despite strenuous opposition from the utility, the group was successful in getting changes to the proposed tariff. When the utility later tried to complicate the process, the group successfully blocked such efforts.

LESSONS LEARNED

Several lessons became obvious during struggles over proposed rate hikes.

- Most PUC proceedings involve the same few talking heads every year. Many commissions are already overburdened with cases involving non-energy issues. Unless prodded, they will continue with "business as usual." Merely raising a few key issues forced new discussion because most of the usual parties did not want them even mentioned.

- When writing and interpreting tariffs, it pays to have many customers simulating in their minds the impacts that specific changes could have on their own facilities. Doing so uncovered a variety of otherwise hidden pitfalls (e.g., simultaneous changes to several parts of the rules can have unexpected effects when combined).

- Rate tariffs are a maze of interlocking rules and legalisms. If you narrow your attention to only the options provided by others, you become just another "rat" in that maze. The ability to innovate is a key to making change.

- When commission staffs get cut or reorganized, one can never be sure of the competency level that results. In one case, a water regulator (lacking any energy background) ended up overseeing changes to complex gas tariffs!

The bottom line here is simple: C&I customers cannot depend on state agencies or the usual rate-setting processes to work in their best interests. Direct involvement is essential, especially during times of rapid economic change.

Investment in such ventures is always speculative, but its rate-of-return compares favorably to energy efficiency and demand-side management measures. While payback periods for the latter may be measured in years, expenses incurred by groups for their rate interventions have sometimes paid back in weeks. Some groups have found that, on average, members may realize over $10 in savings or avoided costs for every dollar of annual dues paid for interventions.

GETTING ORGANIZED

As with most efforts, the first steps are often the toughest. Essential to creating any group is a list of customers having a common interest in making change.

A good resource for determining the largest customers in a utility territory is FERC Form 566. Via that publicly available form, utilities are required to list (on a regular basis) their 20 largest electric and gas customers. Because determining the "top 20" based on revenue or usage may be difficult when customers have many accounts, some utilities err on the side of caution and may list up to 50 of their largest customers on the form.

Form 566 is filed online at https://elibrary.ferc.gov/idm-ws/search/fercgensearch.asp. Choose a date range of 1 year, and check off "Electric" (or gas) in the "Library" options. In "Class/Type Info," choose "Form 566" and, In "Text Search," enter your utility's name.

But don't stop at the top 20. It is an axiom among professional organizers that groups never remain the same size: they either shrink or grow (and you don't want your group to shrink). One of the best places to look for new members is among professional societies and trade associations, especially those representing facility operations and real estate management, such as local chapters of:

• Building Owners and Managers Association (BOMA)

• Institute for Real Estate Management (IREM), and local real estate boards

- Association for Facilities Engineering (AFE)
- International Association of Corporate Real Estate Executives (NACORE)
- International Facility Managers Association (IFMA)
- American Institute of Professional Engineers (AIPE)
- chambers of commerce.

A good presentation made to such groups may be just the trick. Contact the local chapters and offer to speak at one of their meetings (they're always looking for speakers).

GETTING CUSTOMERS TO THE FIRST MEETING

Summoning strangers to a meeting is easier when a near-term issue of common interest is the focus of the initial call. Few customers are willing to commit their time and attention to forming an organization, but most are willing to listen to a specific proposal for action. The topic at hand could be ways to improve utility billing (e.g., fixing a recurring problem), a pending rate case, or a proposal for a new utility service (e.g., providing free interval data to customers). The focus of the first meeting should be a proposal for united action to make members' voices heard, such as a joint letter to the utility and PUC, accompanied by a press release. From such small first steps will come both new members and other successful efforts.

PERSUADING CUSTOMERS TO JOIN

"Selling" membership can also be difficult. Some will come to a meeting and say that "you can't fight City Hall," so why waste time and money trying to do so. The key to overcoming such negativism is to realize that such people are really just looking for someone else to take the responsibility if an effort fails. Many come to meetings to stay current on what their peers are up to, just in case success is a possibility. Understand that if they truly believed that all efforts were futile, they would not have come to your meeting! To handle naysayers, it's often best to acknowledge that there is always the risk of failure—but while doing nothing avoids failure, it also ensures there can never be a success. Sharing the risk among many limits the losses any one member will incur if an effort fails.

It also pays to present the positive aspects of working together. Groups are, for example, instructive for customers new to energy market issues, and are a good way to informally network with others in their industry. Some have found better jobs as a result of this friendly communication.

MEMBERSHIP ISSUES

Credibility is essential to growing and maintaining a customer group. That means choosing wisely who may join and how they may participate.

Vendors, for examples, see any organization of customers as fertile territory to promote their products. Limiting their membership is necessary to prevent meetings from becoming a feeding ground for hungry contractors.

Since most groups are local or state-based, they typically see their utilities as potential adversaries. As a result, credibility demands that utilities be barred from having membership in a customer-controlled group, or at least on its governing board. Failing to follow this credo has resulted in several groups disbanding, making some utilities quite happy with the small investment they made in their dues.

Some customers (especially nonprofit organizations) may be reluctant to lend their names to a group that may participate in a controversy, regardless of how lucrative the outcome could be. Such organizations often depend on donations from utilities, and such funding could be jeopardized if those utilities become displeased. Many nonprofits are, however, large energy consumers and benefit from limiting rate hikes. To allow access by seriously interested consultants and vendors, while not scaring away those who do wish to maintain a low profile, some groups have created different types of membership to address those concerns. Some members remain "private," meaning they may attend meetings but are never identified as belonging to the group.

Members may be invited to bring guests to meetings but are expected to identify them during introductions. Limiting attendance creates a comfort level to other members during sensitive discussions, and demonstrates that the benefits of the group are not open to those who do not financially support the organization. Doing so may also tantalize others wishing to know what's going on that they can't see.

STRUCTURE AND FINANCING

Many groups are constituted as non-profit organizations, but dues and contributions are typically not considered tax-deductible to avoid any conflict with the IRS over lobbying (note that actions before PUCs are not considered lobbying, but actions related to legislation may be). Some groups are registered with the IRS and hold bank accounts and tax ID numbers in their names.

While structures often differ, some form of steering committee is typically elected by the voting membership and may in turn elect officers (Chair, Secretary, Treasurer) from its members. A typical group's bylaws are like those of most professional societies, and can be amended by the membership. While not essential, some rotate their presidents or chairmen each year or two to encourage sharing of responsibility and fresh approaches.

As with any organization, a great deal of attention is paid to collecting and disbursing funds. Various ways to raise funds exist. It is typical, however, for those with the larger energy bills to pay more than those who would benefit less from the group's activities.

Set a realistic dues level, and then work hard to increase

revenue by increasing membership, not by asking the same few to bear a greater burden.

Industrial, commercial, and institutional facilities may use energy very differently, so it pays to query how each facility is operated. Carefully consider, for example, how to bill cogenerators that turn gas into electricity (and thus have a small electric bill) but may require more legal assistance (e.g., with backup/standby rates) than other members.

A potential member may ask why he should pay dues when the results could benefit other customers, including his competition. Focus on the benefit it will have for his rates, and point out how prior actions of the group may have already saved him money. Conclude by pointing out that, if the group fails due to a lack of resources, neither he nor any other customer will get any benefits. It doesn't hurt to add that his small dues contribution may be the only way he will ever "get back at the utility" for the high bills he pays.

ORGANIZATIONAL AND OPERATING RESOURCES

Most of the funds collected by customer groups are used for legal interventions and educational efforts. Few have any paid staff (though many use consultants under contract). Work may also be distributed among steering committee and sub-committee members, who may in turn use their office staff or company resources to handle tasks (e.g., mailings, setting up meetings, etc.). The group's consultant and/or attorney usually keeps the group informed of PUC actions, and the steering committee sets policy, develops meeting agendas, and oversees finances.

Using members' corporate resources may be an excellent way to stretch a group's budget. Members may be willing to "volunteer" their firm's legal staff to review bylaws, accountants to check IRS filings, etc. In some cases, support may be obtained from other organizations having an interest in the same goal, but whose primary focus is elsewhere (e.g., associations of health care facilities, or chambers of commerce). This also holds true for other groups in your state or area. When it comes to controlling costs, there is no room for pride: if another organization's members will benefit from your work, feel free to ask for its help.

Members need to know that their money is being spent wisely—but that it is being spent. The greatest expenses will typically be connected to interventions, which may be quite costly. Total budget for contesting a major rate case is typically measured in five or six figures. By comparison, however, members as a group may save millions of dollars due to this effort. To properly share that burden, expanding membership is essential and should be every member's duty.

KEEPING MEMBERSHIP/ATTENDANCE ATTRACTIVE

Meetings should never be held simply because they are on a schedule. Many groups meet only quarterly or every other month, while their steering committees meet on a monthly or as-needed basis. Each members' meeting has either a speaker or an important item demanding membership attention. It's best to limit meeting length (especially during work hours) to 90 minutes and keep them moving at a brisk pace.

A newsletter is also essential to keep members up-to-date and as a means to attract new members. A private Facebook page and e-mailed briefs are excellent ways to keep all members abreast of issues and fast-breaking events.

For customers new to buying energy, access to expertise may be more immediately important than involvement in interventions. To handle such requests, some groups maintain files of basic contracts and formats (such as for requests for proposals) for their members. Likewise, access to a qualified consultant paid by the group can often save a customer the cost of hiring one. In most cases, one good tip is worth several years' dues, making attendance a good time investment. Keeping such communications closed to outsiders enhances the value of membership.

TIPS FOR INTERVENORS

Beware of focusing only on PUC proceedings. In some cases, the appropriate arena may be elsewhere. Others include: appearing before a legislative committee where PUC policy and actions are reviewed, a governor's office when PUC commissioners are about to be chosen or reappointed, or a court wherein a PUC, utility, ISO, or government agency is challenged. Sometimes, merely filing a lawsuit creates sufficient leverage to bring a stubborn party to the table.

Patience and persistence are very important to winning. Unlike most facility customers, whose focus is on daily building operations, utilities are well equipped to maneuver the often frustrating maze of PUC proceedings. As a result, the process itself is sometimes the desired result: eliminating a foe does not necessarily mean defeating him, when all one need do is to make him lose interest in the battle.

To successfully change an existing tariff or utility rule, it pays to always be looking into the future at the next rate proceeding and developing a plan on what needs to be changed. Doing so will direct you toward what documentation on past (or current) problems must be collected. In one case, a group won changes through a rate proceeding and then got them retroactively applied to an outstanding penalty case, saving nearly a million dollars for its members.

It rarely hurts to be bold. PUCs have complaint procedures for customers who feel they have not been properly treated by a utility, and some utilities hate dealing with them. A complaint from a customer group carries more clout than one from a single customer, and is a good subject for a press release.

Recall that PUC commissioners are not immune from criticism: they are nominated for defined periods, usually by governors; some are elected. It pays to maintain a dossier of commissioner actions your group may consider unacceptable and then use it when an offending commissioner's term is up for renewal.

Always be looking for other resources to expand your group's influence. Often forming liaisons with existing groups is better than duplicating their efforts, while in other cases absorbing a dormant or less active organization is beneficial to both groups.

The "devil is in the details" of tariffs. Whenever a member finds a problem with the way he is being served, it pays to identify the exact line in the tariff or procedure that creates the difficulty. In one case, a utility incorrectly developed a maximum allowable quantity of gas a customer could take in one day, regardless of weather conditions. That quantity was then used to set penalty limits in an unrelated area. By successfully challenging the original error, the group was also able to lower the penalties.

Exercise control when using attorneys and consultants. In the middle of a case, some may lose sight of the limited resources of their clients. Others may see a way to gain professional standing from a decision and use the client more as a sponsor (instead of the beneficiary) of their services. Close financial control is difficult, but essential. Limiting the attorney or consultant to review, advisory and verbal presentation tasks—instead of the entire spectrum of writing, representing, and policymaking—can help to maintain the necessary control in an amicable fashion.

In effect, it is as important to "unbundle" a professional's services and skills as it is to unbundle the utility's services. Otherwise, you may find it difficult to separate the work you really need from the totality for which you are charged. Merely setting a spending cap is poor practice and, in the thick of battle, is not enforceable. A specification of professional services developed prior to the next intervention will instead help ensure budgetary control.

CONCLUSION

Many organizations exist to help businesses and institutions, but few of them are focused on energy costs. As a result, opportunities for cutting such expenses are often lost. Establishing a forum for energy users that combines their interests and talents toward mutual benefit can be a powerful and rewarding venture for all involved.

Glossary of Tariff and Energy Procurement Terms

Terms and definitions seen below were obtained from various uncopyrighted sources that include: proceedings and glossaries from agencies such as FERC, DOE, EIA, EPA, IEA, NERC, various state public utility and other regulatory commissions, regulations fostered by those agencies, professional associations (e.g., IEEE, EEI, NARUC), the Power Marketers Association (PMA), as well as general trade jargon, and the author's own descriptions.

5x16 Market—Typical wholesale pricing designation for power supplied at 100% load factor for the 5 weekdays, between 6 AM and 10 PM (16 hours).

7x24 Market—Typical wholesale pricing designation for power supplied at 100% load factor for the 7 days of the week, during all 24 hours of the day.

Abandonment—Process under which power and gas transmission owners may cease to use assets (e.g., old pipelines). Owners may be required to get FERC permission before abandoning or discontinuing services of facilities under its jurisdiction.

Above-market Cost—The cost of a good or service over that of comparable goods and services in the market. It has often been applied to the cost of a public good or stranded benefit (e.g., electricity) that exceeds or would increase the short-term marginal cost of delivered power.

Access—The contracted right to use an electrical system to transfer electrical energy, e.g., "retail access" is another term for the right of retail customers to receive power supply from non-utility sources.

Access Charge—A charge levied on a power supplier, or its customer, for access to a utility's transmission or distribution system, ostensibly covering that system owner's costs (plus profits) to send someone else's electricity over his wires.

Accumulate—The addition of contracts to a market position.

ACI—See Activated Carbon Injection

Acid rain—Also called acid precipitation or acid deposition, acid rain is precipitation containing harmful amounts of nitric and sulfuric acids formed primarily by sulfur dioxide and nitrogen oxides released into the atmosphere when fossil fuels are burned. It can be wet precipitation (rain, snow, or fog) or dry precipitation (absorbed gaseous and particulate matter, aerosol particles or dust). Acid rain may have a pH below 5.6, which is that of normal rain. A pH measurement of 7 is regarded as neutral, with higher numbers being more alkaline and lower numbers being more acidic.

Activated Carbon Injection (ACI)—Technology for removing mercury from emissions by coal-fired power plants.

Active Demand—European term for "demand response."

Actuals—Physical commodities. Sometimes called a spot commodity or a cash commodity.

Adequacy—See Reliability

Adequacy (electric)—The ability of the electric system to supply the aggregate electrical demand and energy requirements of the end-use customers at all times, taking into account scheduled and reasonably expected unscheduled outages of system elements.

Adequate Regulating Margin—The minimum on-line capacity that can be increased or decreased to allow the electric system to respond to all reasonable instantaneous demand changes to be in compliance with the control performance criteria.

Adjacent System or Adjacent Control Area—Any system or Control Area either directly interconnected with or electrically close to (so as to be significantly affected by the existence of) another system or Control Area.

Adjustment bid—A bid auction conducted by the independent system operator or power exchange to redirect supply or demand of electricity when congestion is anticipated

Administrative Law Judge (ALJ)—Overseer of a rate proceeding; acting as the arbiter of disputes and guarantor that proper procedures are followed.

ADP—See Alternative Delivery Procedure

Affinity Group—Any organization of similar energy customers having a pre-existing relationship that is used to foster aggregated purchasing of energy (e.g., BOMA chapter, Dry Cleaners Association, Industrial Development Group, etc.)

Affordability Programs—Any of a number of utility programs to render utility bills affordable, especially for low-income customers. Among them have been free energy efficiency programs, reduced fuel costs, donations to charitable organizations, energy pricing based on income, forgiveness of late fees or past bills, and/or delay or avoidance of disconnections during extreme weather.

AGA—American Gas Association

AGC—See Automatic Generation Control

Aggregate Net Metering (ANM)—regulatory policy allowing multiple customers to aggregate their loads, and on-site power outputs, to take advantage of net metering capabilities previously limited to individual accounts. First US occurrence: Maryland in November 2013. May also be called "remote net metering."

Aggregator—An entity that assembles customers into a buying group for the purchase of a commodity service. Vertically integrated investor-owned utilities, municipal utilities and rural electric cooperatives perform this function. Buyer cooperatives, power marketers, affinity groups, or brokers are performing this function in restructured power markets.

Aggregators of Retail Customers (ARC)—A demand response (DR) company that gathers curtailable end user load (and/or on-site generating capacity) for purposes of bidding the total into the wholesale DR market (this term not used for energy procurement).

Allowances—The discounts (or premiums) allowed for grades or locations of a commodity lower (or higher) than the par or a basis grade or location specified in the futures contract; also called differentials.

Alternative Delivery Procedure (ADP)—A provision of a futures contract that allows buyers and sellers to make and take delivery under terms or conditions that differ from those prescribed in the standard futures contract. An ADP occurs following termination of futures trading in the spot month and after all long and short futures positions have been matched for the purpose of delivery.

Alternative Retail Electricity Supplier (ARES)—The Illinois name for an unregulated retail electricity supplier. Same as Competitive Retail Energy Supplier (CRES) in Ohio, Retail Electricity Supplier (REP) in Texas, and Energy Services Company ESCo) in New York.

Alternative Technology Regulation Resource (ATRR)—Types of equipment that includes chemical and flywheel batteries designed to dispatch capacity and power like a generator fur use in grid frequency regulation and power support. Sometimes called "limited energy resources" because they are based on stored energy as versus continuously supplied energy, e.g., natural gas or coal.

American Option—An options contract that may be exercised at any time prior to expiration. All New York Mercantile Exchange (NYMEX) energy options are "American." See also European Option.

Ampere—The unit of measurement of electrical current produced in a circuit by 1 volt acting through a resistance of 1 ohm.

AMR—See Automated Meter Reading

AMRA—See Automatic Meter Reading Association

Ancillary Services—Interconnected Operations Services identified by the U.S. Federal Energy Regulatory Commission (Order No. 888 issued April 24, 1996) as necessary to ensure the reliable operation of the transmission system and facilitate power transfers. Services that ensure reliability and support the transmission of electricity from generation sites to customer loads. Such services may include load regulation, spinning reserve, non-spinning reserve, replacement reserve, black start and voltage support.

API—American Petroleum Institute

APPA—The American Public Power Association is a trade association representing the interests of municipal utilities. See also Public Utility.

Apparent power—The product of the voltage (in volts) and the current (in amperes). It comprises both active and reactive power. It is measured in "volt-amperes" and often expressed in "kilovolt-amperes" (kVA) or "megavolt-amperes" (MVA). See Power, Reactive Power, Real Power.

Arbitrage—The simultaneous purchase and sale of similar or identical commodities in two different markets in hope of gaining a profit from price differences.

ARC—See Aggregators of Retail Customers

Area Control Error—The instantaneous difference between actual and scheduled power interchange between two points, taking into account the effects of frequency bias.

ARES—See Alternative Retail Electricity Supplier

ARR—See Auction Revenue Rights

Arrearage—Money owed on past bills.

Ash—Impurities consisting of silica, iron, alumina, and other noncombustible matter that are contained in coal. Ash increases the weight of coal, adds to the cost of handling, and can affect its burning characteristics Ash content is measured as a percent by weight of coal on an "as received" or a "dry" (moisture-free, usually part of a laboratory analysis) basis.

Ask—The lowest price offered at which a security or commodity is available for sale. Also a motion to sell a financial product at a specified price sometimes called an offer.

Ask Size—The number of contracts offered at the ask price (e.g., 150 MW that is being offered in the form of standard 50 MW contracts translates to an ask size of 3 contracts).

Asset—An economic revenue, tangible or intangible, which is expected to provide benefits to a business.

Assignment—The process through which an option seller is notified of the buyer's intention to exercise the rights associated with the option.

At the market—An order placed "at the market" is done immediately at the best price available when it reaches the trading floor.

ATC—Around The Clock. Typically used to describe wholesale electricity pricing that applies to power provided during the entire 24-hour period, 7 days a week (unlike power priced for an on-peak or off-peak period). Same as 'Round the Clock (RTC).

ATC—Available Transmission Capacity. ATC can also mean Available Transfer Capability.

At-the-money—An option whose exercise, or strike, price is closest to the futures price.

Auction—Any of several wholesale market mechanisms that settle pricing by bids from participating parties; different sets of rules apply to different types of auctions. Note that terminology describing auctions is not always consistent from one industry to another.

Ascending Block—An auction in which bidders name how many blocks of power (e.g., 25 MW) they want and may increase that number, but cannot decrease it during the auction. As prices increase, some bidders are expected to drop out, until the time expires.

Declining Block—An auction in which suppliers present their highest bids, and a buyer reduces the price he will accept for additional blocks of power in successive rounds until some suppliers drop out, or reduce the amount of power they are willing to provide. When a buyer has sufficient supplies, it stops its auction, and the price at that point becomes the clearing price for all power.

Descending Clock—Similar to a Declining Block Auction, but time-based:

Three days before the beginning of auction, The Auction Manager informs registered bidders of each utility's starting price (the price in round 1 of the auction) and load caps.

In each round of bidding, bidders provide the number of tranches of load they are willing to serve for each utility at the announced prices (subject to load caps).

If the number of tranches bid is greater than the number of tranches needed by a utility, the prices are reduced in the next round.

In each round, bidders can move tranches between utilities (subject to load caps), but cannot increase total tranches offered statewide from previous rounds.

Prices 'tick down' in each round in decrements pre-determined by a specific formula representing a percentage of the previous price. The auction closes when the supply bid is just equal to the load to be procured.

Bidders holding the final bids when the Auction closes are the winners.

After the descending clock auction is over:

All bidders that win receive the same price per kWh of load served by the utility during the contract term.

Auction Types:

Double—A process of buying and selling goods whereby potential buyers submit their bids and potential sellers simultaneously submit their ask prices in an auction. An auction overseer then chooses a specific price that clears the market and all the sellers who asked less than the clearing price sell and all buyers who bid more than the clearing price buy at this price.

English—A typical auction such as eBay or other arrangement wherein open outcry allows all to be aware of the present bidding price, and the highest bidder wins, taking an entire product or lot.

Classic Dutch (or Reverse)—A high price is set and gradually lowered until a bidder accepts it and pays that amount, for either an entire lot or a portion thereof.

Dutch (or Vickrey)—The highest bidders win, but all pay the lowest winning price; this is essentially the way that most power markets work. The lowest price that provides sufficient capacity to meet demand sets the winning price, with all generators being paid the same price for their power.

Japanese—Very similar to playing poker, in which no new bidders (players) are allowed to bid once bidding (playing) has started, and each bidder must either up his bid to match that of the highest offer or else drop out of the auction (fold).

Auction Revenue Rights (ARR)—A holder of financial transmission rights (FTR) has the right to collect or the obligation to pay congestion rents associated with transmission between specified points of injection and withdrawal, while ARRs are the rights to collect on FTRs.

Automated Meter Reading (AMR)—A form of advanced (or enhanced) metering that uses communications devices to communicate data from the meter to the utility or (in an unbundled metering services environment) the meter reading or meter data management provider. AMR may be used to transmit simple energy usage data from the meter, to transmit more complex measures of energy recorded in the meter, or to implement advanced functionality such as outage detection, remote programming of meters by an authorized party, or other functionality. See also NMR (Network Meter Reading).

Automatic Generation Control (AGC)—Equipment that automatically adjusts a Control Area's generation to maintain its interchange schedule, plus its share of frequency regulation. The following AGC modes are typically available:

Tie Line Bias Control—Automatic generation control with both frequency and interchange terms of Area Control Error considered.

Constant Frequency (Flat Frequency) Control—Automatic generation control with the interchange term of Area Control Error ignored.

Generation Control mode attempts to maintain the desired frequency without regard to interchange.

Constant Net Interchange (Flat Tie Line) Control—Automatic generation control with the frequency term of Area Control Error ignored. This mode attempts to maintain interchange at the desired level without regard to frequency.

Automatic Meter Reading Association (AMRA)—Industry organization for utility metering firms. In 2007, changed name to Utilimetrics, and was managed by Utilimetrics Technology Association. It was disbanded in December 2012.

Availability—A measure of time a generating unit, transmission line, or other facility is capable of providing service, whether or not it actually is in service. Typically, this measure is expressed as a percent available for the period under consideration, (e.g., an availability of 80% typically means that a plant could have produced power for about 7,000 hours during a year (80% x 8,760 hr/yr = 7,008) had requests been made for such service).

Available but not Needed Capability—Net capability of main generating units that are operable but not considered necessary to carry load, and cannot be connected to load within 30 minutes.

Available Margin—The difference between available resources and Net Internal Demand, expressed as a percent of available resources. This is the capacity available to cover random factors such as forced outages of generating equipment, demand forecast errors, weather extremes, and capacity service schedule slippages.

Available Resource—The sum of existing generating capacity, plus new units scheduled for service, plus the net of equivalent firm capacity purchases and sales, less existing capacity unavailable due to planned outages.

Available Transfer Capability (ATC)—A measure of the transfer capability remaining in the physical transmission network for further commercial activity over and above already committed uses. ATC is defined as the Total Transfer Capability (TTC), less the Transmission Reliability Margin (TRM), less the sum of existing transmission commitments (which includes retail customer service) and the Capacity Benefit Margin (CBM).

Average Cost—The revenue requirement (which includes any guaranteed rate-of-return) of a utility divided by the utility's sales. Average cost typically includes the costs of existing power plants, transmission, and distribution lines, and other facilities used by a utility to serve its customers. It also includes operating and maintenance, tax, and fuel expenses.

Average Revenue per Kilowatt-hour—The average revenue per kilowatt-hour of electricity sold by sector (residential, commercial, industrial, or other) and geographic area (State, Census division, and National), is calculated by dividing the total monthly revenue by the corresponding total monthly sales for each sector and geographic area.

Averch-Johnson Effect—The incentive for regulated companies under rate-of-return regulation to overinvest as a way to increase profits.

Avoided Cost—The cost the utility would incur but for the existence of an independent generator or other energy service option. Avoided cost rates have been used as the power purchase price utilities offer independent suppliers known as qualifying facilities. See Qualifying Facility.

AVR—Automatic voltage regulation

BA—See Balancing Authority

Back Out Rate—A credit on an electric bill that purports to represent the cost of electric supply (meaning production, but not transmission or distribution) from the utility. Sometimes also called Supply Credit or Shopping Credit, such numbers are used as a benchmark against which customers seek prices below the credit. That difference is then the amount saved (relative to continued purchasing from the utility) by buying power from a competitive supplier.

Backout Credit—See Back Out Rate

Back-to-Back Trading—See Round Trip

Backup Power—Power provided by contract to a customer when that customer's normal source of power is not available.

Backup Supply Service—See Interconnected Operations Services

Backwardation—A trend in which futures prices are progressively lower in the distant delivery months. Backwardation is the opposite of contango. It is also known as an Inverted Market.

Balance Point Temperature—Temperature at which a building does not need to be mechanically heated or cooled.

Balancing and Settlement—A process of reconciling the total kWh (including Losses during delivery) provided to the system on behalf of a customer (or group of customers) to the amount of power used by that customer during a given time period. Settlement will be completed when the cost of the imbalance between scheduled delivery and aggregated customer usage is assessed. The settlement process relies on readings from interval meters, readings from billing meters, and load profiles to calculate the customer's usage.

Balancing Authority (BA)—A reliability entity responsible for balancing demand and supply within the metered boundaries of its area to support interconnection frequency.

Banking—Energy delivered or received by a utility with the intent that it will be returned in kind in the future. See Storage, Energy Exchange.

Barrel—A volumetric unit of measure for crude oil and petroleum products equivalent to 42 U.S. gallons.

Base Bill—A charge calculated through multiplication of the rate from the appropriate electric rate schedule by the level of consumption.

Baseload—The minimum amount of electric power delivered or required over a given period of time at a constant rate. On an hourly or monthly load profile, this is the level of demand or usage that is seen as a minimum on most hours, including evenings, thereby forming the "base" that peaks rest on.

Baseload Capacity—The generating equipment normally operated to serve loads on an around-the-clock basis.

Baseload Plant—A plant normally operated to take all or part of the minimum load of a system, and which consequently produces electricity at an essentially round-the-clock constant rate. These units are operated to maximize system mechanical and thermal efficiency and minimize system operating costs. Nuclear plants are commonly baseloaded units, for example, due to their very low running (as versus installation) costs.

Basis—The difference that exists at any time between the futures price for a given commodity and the comparable cash or spot price for the commodity. This term may also be used to describe Basis Spread.

Basis grade—The grade of a commodity used as the standard of a contract.

Basis risk—The uncertainty regarding how the spread between cash and futures prices will widen or narrow between the time a hedge position is implemented and liquidated. This term may also be applied to variations in transmission pricing between two points, such as moving natural gas in an interstate pipeline. See also Hedge ratio.

Basis Spread—Differential between the prices of a commodity at two different locations.

BATNA—See Best Alternative To a Negotiated Agreement.

Bbl—Abbreviation for barrel (which is 42 U.S. gallons).

Bcf—Abbreviation for 1 billion cubic feet, usually applied to natural gas.

Bear—One who believes the market is headed lower; also, a down-trending market.

Bear spread—An options position comprised of long and short options of the same type, either calls or puts, designed to be profitable in a declining market. An option with a lower strike price is sold and one with a higher strike price is bought.

Behind The Meter (BTM)—Anything on the customer side of an electric meter (e.g., power production or storage) that may be of concern to the utility and/or grid.

Benefit Cost Analysis (BCA)—A calculation showing the relative financial merits and expenses related to a change in equipment and/or policy.

Best Alternative To a Negotiated Agreement (BATNA)—An option that must be known and potentially feasible in order to secure an acceptable agreement. An opponent must know his adversary has a potent BATNA to which he may revert if the opponent fails to agree to a mutually acceptable arrangement.

Beta—A coefficient that is a measurement of a security's price fluctuation due to factors that simultaneously affect the prices of all marketable securities. A Beta of 1.0 means that the security's price fluctuation is the same as the overall market as measured by the S&P 500 Index. A Beta greater than 1.0 means that a security's price fluctuation is greater than the S&P 500 Index.

Bid—A motion to buy a futures contract at a specified price, or the highest price a prospective buyer is prepared to pay (or the lowest a supplier will offer) for a security or commodity at a particular time.

Bid Size—The number of wholesale contracts offered at the bid price.

Biddable Franchise—See Competitive Franchise

Bilateral agreement—A written statement signed by two parties that specifies the terms for exchanging or buying energy.

Bilateral Contract—A direct contract between the power producer and user or broker outside of a centralized power pool or exchange.

Bilateral Transaction (BT)—A transaction between two willing parties who enter into a physical or financial agreement to trade energy commodities. Bilateral transactions entail reciprocal obligations and can involve direct negotiations or deals made through brokers.

Billing Interval Meter (BIM)—A revenue-grade interval meter used for utility billing.

Billing Unit Effect—FERC term for the net impact that a demand response may have on the power price paid by wholesale customers, once both demand and volume reductions (which may slightly cancel each other) are taken into account. The concept is based on the notion that a reduction in kWh volume due to a demand response then forces fixed costs to be spread across fewer kWh, thus raising their unit cost, even as an incremental cut in demand lowers kWh pricing along an hourly supply-demand price curve.

BIM—See Billing Interval Meter

Biomass—Organic non-fossil material of biological origin constituting a renewable energy source, such as forest waste.

Biomethane—Natural gas originating from landfills, biomass, or anaerobic digesters that is pumped into the natural gas pipeline system rather than being used on-site to generate energy.

Bituminous coal—A dense coal, usually black, sometimes dark brown, often with well-defined bands of bright and dull material, used primarily as fuel in steam-electric power generation, with substantial quantities also used for heat and power applications in manufacturing and to make coke. Bituminous coal is the most abundant coal in active U.S. mining regions. Its moisture content usually is less than 20 percent. The heat content of bituminous coal ranges from 21 to 30 million Btu per ton on a moist, mineral-matter-free basis. The heat content of bituminous coal consumed in the United States averages 24 million Btu per ton, on the as-received basis (i.e., containing both inherent moisture and mineral matter).

Black Start Capability—The ability of a generating unit or station to go from a shutdown condition to an operating condition and start delivering power independently from the rest of the electric system.

Black-Scholes model—An options pricing formula initially derived by Fisher Black and Myron Scholes for securities options and later refined for options on futures.

Blend and extend (B&E)—Procurement technique through which a term fixed price contract is extended to take advantage of pricing in a backwardated forward curve; the curve's pricing of future energy for the extended term is then blended (weighted by projected monthly consumptions) with the pricing in the existing contract to bring down its immediate pricing.

Block Forward—Spot energy destined for future delivery at a specified hour or month; they differ from futures in that they are available only up to 6 months in advance and are intended for physical delivery rather than as a purely financial instrument.

Board of trade—Any exchange or association engaged in the business of buying or selling any commodity or receiving the same for sale on consignment.

Board order—An order that becomes a market order when a strike price is reached. A "sell" order is placed above the market while a "buy" order is placed below the market.

Boiler—A device for generating steam for power, processing, or heating or for producing hot water for heating or hot water supply. Heat from a contained combustion source (typically called the "burner") is transferred to a fluid (typically water) circulating inside the boiler.

Bonneville Power Administration (BPA)—One of five federal Power Marketing Administrations that sell low-cost electric power produced by federal hydroelectric dams to agricultural and municipal users. BPA serves Idaho, Oregon, and Washington as well as parts of Nevada and Wyoming.

Book—When used as a noun, the compiled actives trades of a firm at a given time; when used as a verb, the act of closing a deal that will become part of a firm's "book."

Book transfer—Transfer of title to buyer without physical movement of product.

Booking the basis—A forward pricing sales arrangement in which a cash price is determined within a specified time, at which a previously agreed basis price is added to the then-current futures quotation.

Bottleneck Facility—A point on the system, such as a transmission line, through which all electricity must pass to get to its intended buyers. If there is limited capacity at this point, some priorities must be developed to decide whose power gets through.

Bottom-up—An approach taken when unbundling utility rates, and the opposite of a top-down approach. Bottom-up implies consideration of all components of a utility rate for possible competition, including transmission, distribution, customer service, metering, marketing, taxes, etc. as well as the energy component (which is generally considered the "top" of the rate). A top-down approach involves an examination in the reverse order and typically results in only the energy component becoming subject to competition, instead of all aspects of the utility rate. See also Load Forecast for unrelated use of Bottom-up and Top-Down.

Bourse—European form of a commodity exchange.

Box spread—An options market arbitrage in which both a bull spread and a bear spread are established for a risk-free profit. One spread includes put options and the other includes calls.

BPA—See Bonneville Power Administration

BPL—See Broadband Power Line

Bragawatts—An Enron term for megawatt-hours traded back and forth between traders to plump up their apparent trading volume. Several power trading newsletters and information services quarterly and annually list power trading volumes and rank the various firms to see who's "ahead." A high position on such a list may be seen as basis for claiming a strong position to encourage purchasing of the firm's stock or providing other forms of investment support. While it was typical for a single unit of power to be traded up to 10 times before delivery as part of normal trading, additional exchanges of power solely to portray fictitious volumes end up masking data on real transactions. See also Round Trip.

Break—A rapid and sharp price decline in pricing.

Breakeven point—The underlying futures price at which a given options strategy is neither profitable nor unprofitable. For call options, it is the strike price plus the premium. For put options, it is the strike price minus the premium.

British Thermal Unit (Btu)—A standard unit for measuring the quantity of heat energy required to raise the temperature of 1 pound of water by 1 degree Fahrenheit.

Broadband Power Line (BPL)—Communication via power lines are used to send and receive high-speed data.

Broker—A third party who establishes a transaction between a seller and a purchaser without taking title to the commodity. A power broker may also aggregate customers and arrange for transmission, firming and other ancillary services as needed.

Brokerage—The fee charged by a broker for execution of a transaction. The fee may be a flat amount or a percentage.

Brownian Motion—When price movement is random and the chances of a trade going up or down in value are the same; also called Random Walk.

BT—See Bilateral transaction

BTM—See Behind The Meter

Btu—See British thermal unit

Bucket—A gathering of items of a similar type, such as trades of a similar value, or a time period during which trades may take place.

Bulk Electric System—A term commonly applied to the portion of an electric utility system that encompasses the electrical generation resources and bulk transmission system.

Bulk Power Supply—Often this term is used interchangeably with wholesale power supply. In broader terms, it refers to the aggregate of electric generating plants, transmission lines, and related equipment. The term may refer to those facilities within one electric utility, or within a group of utilities in which the transmission lines are interconnected.

Bull—One who believes that prices are headed higher; also, an up-trending market.

Bull spread—An options position composed of both long and short options of the same type, either calls or puts, designed to be profitable in a rising market. An option with a lower strike price is bought and one with a higher strike price is sold.

Bundled utility service (electric)—A means of operation whereby energy, transmission, and distribution services, as well as ancillary and retail services, are provided by one entity.

Buy in—Making purchase to cover a previous sale; often called "covering."

Buy Through—An agreement between utility and customer to import power when the customer's service would otherwise be interrupted and to price such power at prevailing spot market rates.

Bypass—A situation that allows a customer to get full or partial electricity from a non-utility supplier instead of a utility, typically by connecting to a main transmission line instead of using the local utility's distribution system.

C/I—Abbreviation for "Commercial and Industrial" customers; also C&I.

Calendar spread—An options position comprised of the purchase and sale of two option contracts of the same type that have the same strike prices but different expiration dates. Also known as a horizontal or time spread.

California power exchange—A State-chartered, non-profit corporation which provided day-ahead and hour-ahead markets for energy and ancillary services in accordance with the power exchange tariff. It was closed in 2001 and essentially succeeded by the California ISO.

Call Center—A telecommunications facility operated by, or on behalf of, an energy services supplier to receive customer inquiries and perform telemarketing tasks.

Call Option—An option which gives the buyer, or holder, the right, but not the obligation, to buy a futures contract at a specific price within a specific period of time in exchange for a one-time premium payment. It obligates the seller, or writer, of the option to sell the underlying futures contract at the designated price, should an option be exercised at that price. See also Put Option.

Candlestick chart—A graphic showing a vertically oriented rectangle (the candle) whose upper and lower bounds represent opening or closing prices of a bidding session, and a thin center line (the wick) protruding upward or downward from the rectangle's top and bottom, representing the highest and lowest bids during a bid session; also called a Japanese candlestick chart.

Capability—The maximum load that a generating unit, generating station, or other electrical apparatus can carry under specified conditions for a given period of time without exceeding approved limits of temperature and stress.

Capacity (Delivery or Receipt)—The amount of electric power delivered or required for which a generator, turbine, transformer, transmission circuit, station, or system is rated by the manufacturer.

Capacity (Generating or Supply)—The rated continuous load-carrying ability, expressed in megawatts (MW) or megavolt-amperes (MVA) of generation, transmission, or other electrical equipment. Types of capacity include:
  Baseload Capacity—Capacity used to serve an essentially constant level of customer demand. Baseload generating units typically operate whenever they are available, and they generally have a capacity factor that is above 60%.
  Firm Capacity—Capacity that is as firm as the seller's native load unless modified by contract. Associated energy may or may not be taken at option of purchaser. Supporting reserve is carried by the seller.
  Installed Capacity—Also called ICAP, the total wattage of all generators able to be scheduled to serve a given service or control area.
  Intermediate Capacity—Capacity intended to operate fewer hours per year than baseload capacity but more than peaking capacity. Typically, such generating units have a capacity factor of 20% to 60%.
  Net Capacity—The maximum capacity (or effective rating), modified for ambient limitations, that a generating unit, power plant, or electric system can sustain over a specified period, less the capacity used to supply the demand of station service or auxiliary needs.
  Peaking Capacity—Capacity used to serve peak demand. Peaking

generating units operate a limited number of hours per year, and their capacity factor is normally less than 20%.

Unconstrained Capacity—Also called UCAP, the total wattage of all generators that actually deliver power to serve a given service or control area. UCAP may be determined through a de-rating process that corrects for loss of capacity due to high air temperatures (which may reduce capacity of combustion turbines), past failure probabilities for specific generators, or other means.

Capacity (Purchased)—The amount of energy and capacity available for purchase from outside the system.

Capacity Benefit Margin (CBM)—That amount of transmission transfer capability reserved by load serving entities to ensure access to generation from interconnected systems to meet generation reliability requirements. Reservation of CBM by a load serving entity allows that entity to reduce its installed generating capacity below that which may otherwise have been necessary without interconnections to meet its generation reliability requirements. See Available Transfer Capability (ATC).

Capacity Charge—An element in a two-part pricing method used in capacity transactions (energy charge is the other element). The capacity charge is assessed on the amount of capacity being used or purchased. In a retail or wholesale tariff, it may be based on the results of scheduled capacity auctions multiplied by a facility's coincident peak with that of the grid serving it.

Capacity Emergency—A state when a system's or pool's operating capacity plus firm purchases from other systems, to the extent available or limited by transfer capability, is inadequate to meet the total of its demand, firm sales, and regulating requirements. See Energy Emergency.

Capacity Factor—The ratio of the total energy actually generated by a generating unit for a specified period to the maximum possible energy it could have generated if operated at the maximum capacity rating for the same specified period, expressed as a percent; not to be confused with availability, which addresses how often that same plant could have generated such energy. See Availability.

Capacity Interconnection Rights (CIR)—Payment to an ISO required to connect a new source of generation to the grid.

Capacity Margin—The difference between net capacity resources and net internal demand expressed as a percent of net capacity resources.

Capacity Release—A secondary market for gas pipeline capacity that is contracted by a customer or offered by a utility which is not using all of its capacity.

Capacity Tag—In markets charging for electric capacity, the kW load of a facility occurring at the time the local grid peaks, typically on a very hot weekday, plus a defined reserve margin (e.g., 10%). Where interval metering does not exist, it may instead be the average of the highest 4 or 5 monthly kW peaks seen during the prior cooling season. In many cases, it occurs at the same time as the facility's annual peak load.

CAPP—Abbreviation for Central Appalachia Coal

Captive Customer—A customer who does not have realistic alternatives to buying power from the local utility, even if that customer had the legal right to buy from competitors.

Carbon Capture and Sequestration (CCS)—Any of several different processes for capturing CO_2 from exhaust due to burning fossil fuels and then sealing it into geologic formations (or using it in commercial applications) instead of releasing it into the atmosphere.

Carrying charge—The total cost of storing a commodity; includes actual storage charges, insurance, interest on loans, and opportunity loss on committed capital.

Cascading—The uncontrolled successive loss of system elements triggered by an incident at any location. Cascading results in widespread service interruption, which cannot be restrained from sequentially spreading beyond an area predetermined by appropriate studies. In the analysis of energy-efficiency measures, the accounting for interactions that may change the impact of measures that follow others; e.g., the calculated savings due purely to the installation of occupancy sensors will be lessened if "cascaded" after a lamp/ballast upgrade that reduces the wattage left to save by automatically shutting off lights in unoccupied rooms.

Cash commodity—The actual physical commodity. Sometimes called a spot commodity or actuals.

Cash Desk—A trading group that handles hourly or daily trades. See also Term Desk.

Cash Market—The market for a cash commodity where the actual physical product is traded.

CAVEE—See Customer Association, Validation, Editing and Estimation

CBM—See Capacity Benefit Margin

CCA—See Community Choice Aggregation

CCS—See Carbon Capture Sequestration

CDR—ERCOT term for Capacity, Demand, and Reserves; a proceeding and document that determines the available margin between power that is available and load that must be served in the coming 2 years.

CEII—See Critical Energy Infrastructure Information

Certificate of Public Convenience and Necessity—A certificate issued by a regulatory body that allows the recipient (e.g., a utility) to build an energy-related installation (e.g., power plant, pipeline) and include its capital cost in the recipient's rate base. It may also be called a Certificate of Need.

Certification—The process of granting permission to do business or to sell a particular product. A device used by states to enforce standards of conduct and quality on competitive suppliers, on pain of penalties including civil fines, suspension of certification, and revocation of certification to do business.

CF/D—Cubic feet per day. Usually used to quantify the rate of flow of a gas well or pipeline.

CfD—See Contracts for Differences

CFO—Chief Financial Officer of a company, often the person whose permission is needed to fund and implement energy-related activities.

CFTC—See Commodity Futures Trading Commission

Chauffage—Process under which a customer purchases the end result of an electrical or mechanical system (e.g., lumens, temperature maintenance) rather than buying or leasing such systems.

CHP—See Combined Heat and Power (CHP) Plant or Cogeneration

Churn—Action during which a single contract changes hands and pricing many times, indicating liquidity; also describes changeover of personnel or contractors within a given firm.

CIF—See Cost, Insurance, Freight

CIR—See Capacity Interconnection Rights

Circuit—A conductor or a system of conductors through which electric current flows.

Circular Scheduling—A wholesale transaction technique for creating false congestion on power lines to drive up the cost of wholesale power for short periods. See also Death Star.

CIS—See Customer Information System—Process and/or software for managing utility customer data.

City Gate—The point where the systems of an interstate pipeline and a local gas distribution company meet. Pricing for natural gas bought from a marketer is often set at a City Gate, to which local distribution (and possibly balancing) charges must be added.

Class of options—All call options, or all put options, exercisable for the same underlying futures contract and which expire on the same expiration date.

CLEAN—Acronym for Clean Local Energy Accessible Now. Essentially a feed-in tariff program aimed at increasing wholesale distributed generation (WDG), primarily renewables.

Clean Power Plan (CPP)—Ill-fated effort under the Obama administration to reduce emissions from the power sector via statutory limits and promotion of energy efficiency, renewables, and natural gas. Blocked by the Trump administration and various court decisions, it was never implemented, though some utilities planned as if it had been.

Clearing member—Member of an exchange. Also known as an exchange member.

Clearinghouse—An exchange-associated body charged with the function of insuring the financial integrity of each trade. Orders are "cleared" by means of the clearinghouse becoming the buyer to all sellers and the seller to all buyers.

Close—The price of the last transaction completed for the commodity trading session.

Closing range—A range of prices at which transactions took place at the closing of the market.

CMH—Ceramic Metal Halide technology, used in high-efficiency lighting, that exploits advances in gases, containment, ballasts, etc., to create small HID lamps (down to 20 watts) that rival incandescent light sources in output and concentration, at a fraction of incandescent wattage. Eventually made obsolete by developments in LED lighting.

CMR—See Customer Relationship Management

CMS—See Congestion Management System; See also Transmission Congestion Contract (TCC), Firm Transmission Rights (FTR), Financial Congestion Rights (FCR)

CMT—Cost management technique

Coal—A readily combustible black or brownish-black rock whose composition, including inherent moisture, consists of more than 50 percent by weight and more than 70 percent by volume of carbonaceous material. It is formed from plant remains that have been compacted, hardened, chemically altered, and metamorphosed by heat and pressure over geologic time.

Coal Combustion Residuals (CCR)—EPA term for the ash created by burning coal.

Coal-Direct Chemical Looping—process that combines coal with heated iron oxide (e.g. rust) to produce a different form of iron oxide, CO_2, water, and heat. ~98% of the CO_2 may then be captured and the iron oxide recycled. This carbon capture system was in test stages in early 2013.

Coal synfuel—Coal-based solid fuel that has been processed by a coal synfuel plant; and coal-based fuels such as briquettes, pellets, or extrusions, which are formed from fresh or recycled coal and binding materials.

Co-firing—simultaneous burning of natural gas and coal in the same utility boiler, especially when using coal with low Btu content, as a way to maintain or boost MW output, or to sidestep limits on non-carbon emissions (SO_x, mercury) without adding scrubbers.

Cogeneration—Production of electricity from steam, heat, or other forms of energy produced as a by-product of another process. Also called Combined Heat and Power (CHP).

Cogenerator—A generating facility that produces electricity and another form of useful thermal energy (such as heat or steam) used for industrial, commercial, heating, or cooling purposes. Many cogeneration systems are also Qualifying Facilities (QF). To receive status as a qualifying facility (QF) under the Public Utility Regulatory Policies Act (PURPA), the facility must produce electric energy and "another form of useful thermal energy through the sequential use of energy," and meet certain ownership, operating, and efficiency criteria established by the Federal Energy Regulatory Commission (FERC). (See the code of Federal Regulations, Title 18, Part 292.)

Coincident Demand—Two or more power demands that occur in the same time interval.

Coincident Peak Load—Two or more peak loads that occur in the same time interval.

Coke (petroleum)—A residue high in carbon content and low in hydrogen that is the final product of thermal decomposition in the condensation process in cracking. This product is reported as marketable coke or catalyst coke. The conversion is 5 barrels (of 42 U.S. gallons each) per short ton. Coke from petroleum has a heating value of 6.024 million Btu per barrel.

Cold ironing—Powering of docked ships by an on-land generator to minimize fuel/electric cost, noise, and pollution, allowing maintenance of shipboard generators.

Collar—A commodity supply contract between a buyer and a seller in which the buyer is assured of a defined maximum price and the seller is assured of a defined minimum price. This is analogous to an options fence, also known as a range forward.

Combined Cycle—An electric generating technology in which electricity and process steam is produced from otherwise lost waste heat exiting from one or more combustion turbines. The exiting heat is routed to a conventional boiler or to a heat recovery steam generator (HRSG) for use by a steam turbine in the production of electricity. This process increases the efficiency of the electric generating unit.

Combined Cycle Unit—An electric generating unit that consists of one or more combustion turbines and one or more boilers with a portion of the required energy input to the boiler(s) provided by the exhaust gas of the combustion turbine(s).

Combined heat and power (CHP) plant—A plant designed to produce both heat and electricity from a single heat source. Note: This term is being used in place of the term "cogenerator" that was used by EIA in the past. CHP better describes the facilities because some of the plants included do not produce heat and power in a sequential fashion and, as a result, do not meet the legal definition of cogeneration specified in the Public Utility Regulatory Policies Act (PURPA).

Combined Pumped-storage Plant—A pumped-storage hydroelectric power plant that uses both pumped water and natural stream flow to produce electricity.

Combustion Turbine—Any of several types of high speed (usually gas-fired) generators using principles and designs of jet engines to produce low cost, high efficiency power.

Commercial—The commercial sector is generally defined as non-manufacturing business establishments, including hotels, motels, restaurants, wholesale businesses, retail stores, and health, social, and educational institutions. A utility may also classify the commercial sector as all consumers whose demand or annual use exceeds some specified limit. The limit may be set by the utility based on the rate schedule of the utility.

Commercial Operation—Commercial operation begins when control of the loading of the generator is turned over to the system dispatcher.

Commercial sector—An energy-consuming sector that consists of service-providing facilities and equipment of businesses; Federal, State, and local governments; and other private and public organizations, such as religious, social, or fraternal groups. The commercial sector includes institutional living quarters. It also includes sewage treatment facilities. Common uses of energy associated with this sector include space heating, water heating, air conditioning, lighting, refrigeration, cooking, and running a wide variety of other equipment. Note: This sector includes generators that produce electricity and / or useful thermal output primarily to support the activities of the above-mentioned commercial establishments.

Commercialization—Programs or activities that increase the value or decrease the cost of integrating new products or services into the electricity sector. See Sustained Orderly Development.

Commission—Used in energy markets as a shorthand reference to The Federal Energy Regulatory Commission (FERC) or a local (usually state) regulatory commission, typically called a Public Utility Commission or PUC. See Federal Energy Regulatory Commission and Public Utility Commission. In futures or retail power transactions, the fee charged by a broker for the execution of an order for a trade or purchase.

Commissioning—Process for verifying (and possibly fixing) proper operation of every component and system related to a building's energy systems at the point that the building is first turned over to its management. Retro-commissioning (RCx) is the same process as applied after the building has been in operation for several years. Continuous (or Ongoing) Commissioning (CCx or OGCx) is the same process as applied on a routine basis instead of several years apart.

Commission House—An organization that buys and sells actual commodities and / or futures contracts for the accounts of its customers in return for a fee. See also Wire House.

Commitment—Total number of contracts in a commodity or options market that are still open, meaning they have not been exercised, closed out, or allowed to expire. Also called "open interest."

Commodity—A standardized product for sale, generally not very differentiable by vendor.

Commodity Cost—The cost of a commodity (natural gas or electricity) and related charges to deliver it to the marketplace.

Commodity Futures Trading Commission (CFTC)—The federal regulatory body which oversees commodity futures trading activities, standards, and practices.

Common carrier—A person or company having state or federal authority to perform public transportation for hire.

Commonly Owned Unit—A generating unit whose capacity is owned or leased and divided among two or more entities. Synonym: Jointly Owned Unit.

Community Choice Aggregation (CCA)—A movement to allow municipalities to aggregate their members (including all rate classes) into a single energy buying group.

Community net metering—Community net metering allows customers that do not otherwise have access to on-site generation, e.g., rental customers, to build and own net metered projects such as solar photovoltaic (PV) systems.

Competitive Franchise—A process whereby a municipality (or group of municipalities) issues a franchise to supply electricity in the community to the winner of a competitive bid process. Such franchises can be for bundled electricity and transmission/distribution, or there can be separate franchises for the supply of electricity services and the transmission and distribution function. Franchises can be, but typically are not, exclusive licenses. Through the terms of the request for proposals and the negotiation of the franchise agreement, the community can seek suppliers willing to provide electricity consistent with local values, such as energy efficiency, renewable resource development, and job creation. See Value-driven Aggregator. Sometimes called a biddable franchise.

Competitive Market—An environment that allows many sellers and buyers to buy and sell goods or services from each other, unfettered by heavy regulation.

Competitive Retail Area (CRA)—region in which retail access exists within an ISO containing still-regulated areas, e.g., southern Illinois is within MISO (most of which is regulated [2016]) but all of that state offers retail access.

Competitive Retail Energy Supplier (CRES)—Ohio term for an unregulated retail electricity supplier. Same as Retail Electricity Supplier (REP) in Texas, Alternative Retail Electric Supplier (ARES) in Illinois, or Energy Services Company (ESCo) in New York.

Competitive Retail Solutions—MISO term for setting up Forward Resource Auctions (FRA) that foster capacity resources 3 years in advance of projected need.

Competitive Transition Charge (CTC)—A non-bypassable charge generally placed on distribution services to recover utility costs incurred as a result of restructuring (e.g., stranded costs—usually associated with generation facilities and services) and not recoverable in other ways. See also Nonbypassable Charge.

Compound Annual Growth Rate (CAGR)—Business term to describe the year-to-year growth rate of an investment across a known time period. It's calculated by taking the nth root of a total percentage rate of growth, where n = no. of years being analyzed.

Compressor Station—A facility that is used to compress natural gas in order to create additional pressure to increase the amount of gas a pipeline can hold, help move it through a pipeline, or to move it into or from storage.

CONE—Cost of New Entry generation, meaning the threshold $/kW cost for capacity if a standard new generator (typically a single-cycle natural gas fired unit) was to be added to an existing grid. Net CONE is an ISO New England term meaning an estimate of the capacity revenue needed by a new generator in its first year of operation to make it economically viable to build a power plant.

Congestion—a condition (also called "line congestion") existing when available capacity of a transmission line is being closely approached (or exceeded) by the power that Load Serving Entities (LSEs) seek to transmit through it. At such times, alternative power line pathways (or local generators near the load) must instead be used. Congestion need not actually be happening: the ISO need only believe that it may occur at a defined time in order to declare a line congested, thus forcing others to resort to alternatives means of power delivery. To create congestion (and thus force up prices and/or force use of more expensive local generation), some marketers have scheduled fictitious load to make it appear that a line will be fully needed or overloaded. They then drop that schedule after others have already arranged for alternative delivery means. They either sell that local power at a higher price (if they had control of it), and/or get paid by an ISO to drop the fictitious load that caused the problem in the first place, thus making money through nothing except chicanery. Enron referred to this general tactic as Death Star or Load Shift, depending on how money was made.

Congestion Management System (CMS)—Any of several methods to control the cost of transmission congestion.

Congestion Revenue Rights (CRRs)—Also called financial transmission rights (FTR), or transmission congestion contracts (TCC)/auctions.

   With an option congestion revenue right (CRR), a purchaser is entitled to revenue if congestion occurs from the electricity source to the destination, but the buyer is not obligated to pay if congestion occurs in the opposite direction.

   In contrast, an obligation CRR would require the buyer to pay for congestion if it occurs in the opposite direction of the CRR's original direction.

Connection—The physical contact point (e.g. transmission lines, transformers, switch gear, etc.) between two electric systems permitting the transfer of electric energy in one or both directions.

Conservation—A reduction in energy consumption that corresponds with a reduction in service demand. Service demand can include buildings-sector end uses such as lighting, refrigeration, and heating; industrial processes; or vehicle transportation. Unlike energy efficiency, which is typically a technological measure, conservation is better associated with behavior. Examples of conservation include adjusting the thermostat to reduce the output of a heating unit, turning off lights or appliances, and car-pooling.

Conservation Voltage Reduction (CVR)—Operation of a grid's distribution system at the low end of an allowable voltage band (e.g., 115 volts within a 114 to 126 range). Doing so saves energy for both the utility and the consumer, but may slightly impact operation of some customer equipment (e.g., microwave oven).

Consolidated Utility Bill (CUB)—In deregulated states, a utility delivery bill that incorporates the charges from a non-utility supplier.

Construction Work In Progress (CWIP)—The charge shown on a utility's balance sheet for construction work not yet completed but in process. This balance line item may or may not be included in the rate base.

Consumption—The amount of fuel used for gross generation, providing standby service, start-up and/or flame stabilization. The quantity of electricity or natural gas consumed during a utility billing period.

Contango Market—A futures market that is deemed "normal" when carrying charges are reflected in proportionately higher prices for increasingly distant futures contracts. The opposite of Backwardation.

Contingency—The unexpected failure or outage of a system component, such as a generator, transmission line, circuit breaker, switch, or other electrical element. A contingency also may include multiple components, which are related by situations leading to simultaneous component outages.

Contingency Order—An order which becomes effective when a set of specified conditions are met.

Contract Documents—Documents that comprise an agreement between two parties to deliver and/or accept a product or service.

Contract Expiration Date—The last date on which the contract can be exercised.

Contract Grade—That grade of product established in the rules of a commodity futures exchange as being suitable for delivery against a futures contract.

Contract High—The highest intra-day price within the life of the contract.

Contract Low—The lowest intra-day price within the life of the contract.

Contract Path—The most direct physical transmission tie between two interconnected entities. When utility systems interchange power, the transfer is presumed to take place across the contract path, notwithstanding the electrical fact that power flow in the network will distribute in accordance with network flow conditions, generally along the path of least resistance. This term can also mean to arrange for power transfer between systems. See also Parallel Path Flow.

Contract Price—Price of fuels marketed on a contract basis covering a period of 1 or more years. Contract prices reflect market conditions at the time the contract was negotiated and therefore remain constant throughout the life of the contract or are adjusted through escalation clauses. Generally, contract prices do not fluctuate widely.

Contract Receipts—Purchases based on a negotiated agreement that generally covers a period of 1 or more years.

Contracts for Differences (CfD)—a bilateral contract with an unhedged fixed price; the seller is paid a fixed amount over time based on a combination of the short-term (e.g., hourly, daily, monthly) market price and a variable adjustment from a fixed price. If the price rises above that variable level, the seller pays to bring it down to the fixed level. If the price drops below the fixed level, the buyer pays the seller the difference. The fixed price is set high enough so that the seller is taking minimal risk. In a retail contract, the seller may also charge a fixed adder to cover capacity, ancillaries, profit, etc.

Note that CfD pricing is set for each account, and not a single price for a batch of accounts.

Control Area—An electric power system or combination of electric power systems in which a common automatic control scheme is applied to: (1) match, at all times, the load in the electric power system(s) by dispatching generators within the electric power system(s) and capacity and energy purchased from entities outside the electric power system(s); (2) maintain, within the limits of Good Utility Practice, scheduled interchange with other Control Areas; (3) maintain the frequency of the electric power system(s) within reasonable limits in accordance with Good Utility Practice; and (4) provide sufficient generating capacity to maintain operating reserves in accordance with Good Utility Practice. Regional Transmission Organizations (such as an ISO) are becoming the mega-control areas of the future.

Convergence—The allying and/or merging of electric and natural gas suppliers/marketers due to common goals and business interests.

Convergence bidding—Also called virtual bidding. A hedging tool at ISOs, used for hedging or speculation, which allows market players to place virtual bids to buy or sell energy in the day-ahead market, at the same time assuming an opposite obligation to sell or buy that amount of energy back in the real-time market. No power is ever delivered under such deals. A virtual bidder gets paid the difference in the day-ahead price and real-time price, or if he bets wrong, he pays the difference.

Conversion—A delta-neutral arbitrage transaction involving a long futures, a long put option, and a short call option. The put and call options have the same strike price and same expiration date.

Co-op—See Cooperative Electric Utility

Cooperative Electric Utility—Often abbreviated to Co-op, any electric utility legally established to be owned by and operated for the benefit of those using its service. Such entities may generate, transmit, and/or distribute supplies of electric energy to a specified area not being serviced by another utility. Such ventures are generally exempt from Federal income tax laws. Most electric cooperatives were initially financed by the Rural Electrification Administration, a division of the U.S. Department of Agriculture. Rural electric cooperatives may generate and purchase wholesale power, arrange for the transmission of that power, and then distribute the power to serve the demand of rural customers.

Coordinated Transaction Scheduling (CTS)—A process through which two ISOs coordinate real-time power flows at interchanges between their territories to ensure that use of low cost sources of power are maximized and use of higher cost sources is minimized. May also be called "tie optimization."

Core Customer—A customer who depends on a utility for the delivery of energy or services. This customer generally does not have alternatives to buy energy from a non-utility supplier.

Cost of Service—The cost claimed by a utility to provide service to its customers.

Cost, Insurance, Freight (CIF)—Term refers to a sale in which the buyer agrees to pay a unit price that includes the FOB value at the port of origin plus all costs of insurance and transportation. This type of transaction differs from a "delivered" agreement in that it is generally exclusive of duties, and the buyer accepts the quantity and quality at the loading port rather than at the unloading port.

Cost-Based Pricing—Prices controlled by the fees charged by a provider (e.g., a utility), as versus market-based pricing.

Cost-Based Rates—A ratemaking concept used for the design and development of rate schedules to ensure that the filed rate schedules recover only the cost of providing the service.

Cost-Based Regulation—Adjustment of rates based on costs rather than market prices.

Cover—To close out a short futures position by purchasing either a physical commodity or futures for it.

Cost-of-service regulation—A traditional electric utility regulation under which a utility is allowed to set rates based on the cost of providing service to customers and the right to earn a limited profit.

Cost-of-service study—An analysis, typically produced by a utility, to support how it allocates its costs to each rate class.

Covered Writing—The sale of an option against an existing position in the underlying futures contract; e.g., short call and long futures.

CRES—See Competitive Retail Energy Supplier

Critical Energy Infrastructure Information (CEII)—Information concerning proposed or existing critical infrastructure (physical or virtual) that: Relates to the production, generation, transmission or distribution of energy; Could be useful to a person planning an attack on critical infrastructure; Is exempt from mandatory disclosure under the Freedom of Information Act; and Gives strategic information beyond the location of the critical infrastructure.

CRRs—See Congestion Revenue Rights

CSAPR—Cross State Air Pollution Rule; 2011 EPA regulations for cutting emissions (especially of NOx, SOx, and fine particulates) arising mostly from coal and oil-fired power plants. Pronounced "casper" by some utility personnel.

CSS—See Carbon Capture and Sequestration

CT—Abbreviation for Combustion Turbine or Current Transformers.

CTC—See Competitive Transition Charge

CTS—See Coordinated Transaction Scheduling

CUB—See Consolidated Utility Bill

Current (Electric)—A flow of electrons in an electrical conductor. The strength or rate of movement of the electricity is measured in amperes.

Current Transformers—Small doughnut-shaped units that fit around power conduits to assist the measurement of the current passing through such power lines.

Curtailability—The right of a transmission provider to interrupt all or part of a transmission service due to constraints that reduce the capability of the transmission network to provide that transmission service. Transmission service is to be curtailed only in cases where system reliability is threatened or emergency conditions exist. May also be called Interruptibility.

Curtailable Load—Any electrical load that may be curtailed with a user's permission in exchange for a financial incentive from either an LDC, ISO, energy vendor, or state agency. Also called Price Responsive or Price Sensitive Load.

Curtailment—A reduction in the scheduled capacity or energy delivery. Also used to describe a reduction in available renewable power generation due to temporary oversupply, often ascribed to transmission and power storage limitations, but also to excess renewable supply, e.g., high hydro output to avoid flooding forces curtailment of wind turbine operation.

Customer Association, Validation, Editing and Estimation (CAVEE)—The process in which meter data, which has been translated into a common format, is processed and readied for bill generation and other operations that require validated usage data. In CAVEE, meter data are associated with the correct customer through reference to the service provider's database. The data is then subjected to a number of tests for reasonableness, compared to stored usage history and transmitted to billing and other appropriate functions. If data is missing, usage estimations are made according to fixed rules. If anomalies are found error reports are generated.

Customer choice—The right of customers to purchase energy from a supplier other than their traditional supplier or from more than one seller in the retail market.

Customer Obligations—The responsibilities that a customer has, such as paying bills, maintaining safety of a natural gas or electric hook-up, and preventing theft of service.

Customer Relationship Management (CMR)—CMR involves finding ways to maximize the value of the contact between supplier and customer, primarily by leveraging that relationship to offer and sell other services—such as telecommunications, Internet access, and building security—and/or brokering that contact with others who wish to access that same customer base.

Customer Service Protections—The rules governing grounds for denial of service, credit determinations, deposit and guarantee practices, meter reading and accuracy, bill contents, billing frequency, billing accuracy, collection practices, notices, grounds for termination of service, termination procedures, rights to reconnection, late charges, disconnection/reconnection fees, access to budget billing and payment arrangements, extreme weather, illness or other vulnerable customer disconnection protections, and the like. In a retail competition model, would include protections against slamming and other hard-sell abuses.

CVR—See Conservation Voltage Reduction

CWIP—See Construction Work In Progress

Daily Limit—Maximum price variation that commodities and options markets are allowed to experience in one day.

DAM—See Day-Ahead Market

Dark spread—The difference between the spot market values of coal and electricity at an given time, based on the conversion efficiency (Btus to produce a kWh) of a coal-fired plant. At a given plant, one might (e.g.) burn 10,000 Btu to make a kWh. If that coal cost (e.g.) \$5/MMBtu, it could make power at \$.05/kWh. If the wholesale grid price of power at that time in the same place was $.07/kWh, the dark spread would be $.02/kWh (i.e., $20/MWhr). As the conversion efficiency becomes greater, the spread between the market value of the coal and that of power derived by burning the coal becomes wider.

Date of Contract High—The date of the lifetime contract high.

Date of Contract Low—The date of the lifetime contract low.

Date of Last Report—For mutual funds, this is the date of the most recent update of net asset value (NAV) provide by NASDAQ. For money market funds this is the most recent yield provided by NASDAQ. For indexes, this is the date of the most recent update.

Day Trade—The purchase and sale of a futures contract on the same day.

Day-Ahead Market (DAM)—Wholesale power pricing provided in 24-hour blocks or pricing by the hour, provided up to 24 hours in advance, for the next 24 hours.

Day-ahead schedule—A schedule prepared by a scheduling coordinator or the independent system operator before the beginning of a trading day. This schedule indicates the levels of generation and demand scheduled for each settlement period that trading day.

Day-ahead/real-time (DA/RT) spread—The difference (in $/MWh) between wholesale pricing in the DAM and HAM markets at a specific point in time. A positive value indicates that day-ahead prices are higher than real-time prices. A negative number means that real-time pricing at that time was higher than day-ahead pricing.

Daylighting—The use of natural light in conjunction with adjustable electric lighting to reduce power used by the electric lighting, typically through the use of automatic light sensing controls (e.g., photosensors).

DBOOM—Design, Build, Own, Operate, Maintain.

DC—See Direct Customer

DD—See Degree-Day

Death Star—An Enron term for taking advantage of transmission congestion. Line congestion (such as between northern and southern California) may be caused by scheduling transmission (i.e., planning to send power) in the direction opposite to that of the congestion. Doing so reduces the congestion, thus earning a trader a payment from the ISO. No energy, however, was actually put on the grid or taken off. Even if power had been provided (possibly by others being paid by the trader to generate) to relieve that congestion, the cost to do so was often much lower than the congestion payment, thus still earning a profit while doing very little. See also Circular Scheduling.

DEC Bid—Short for decremental bid; a DEC bid is submitted by a generator to the ISO during congestion situations to determine how much he should be paid to reduce the output of a generation unit. The ISO then pays the generator to decrease the amount of energy his plant is producing on one side of a congested point at the same time it pays less-efficient units on the other side of the congested point to increase output.

Deficiency Charge—See ICAP Deficiency Charge.

Degree-Day (DD)—A unit describing a temperature difference, maintained over a 24-hour period, of 1 degree Fahrenheit (1°F). Degree-days are typically calculated as occurring between the average dry bulb outdoor temperature and a base dry bulb indoor temperature (typically 65°F). For example, an average outdoor temperature of 40°F occurring across a 24-hour period would result in 65°F—40°F = 25 degree-days. When occurring below that base temperature, such degree-days are called "heating degree-days" (HDD). When occurring above that base temperature, such degree-days are called "cooling degree-days" (CDD).

Deintegration—See Disaggregation

Deliverability—The amount of natural gas that a well, production field, pipeline, or distribution system can deliver in a given period.

Delivering Party—The entity supplying the capacity and/or energy to be transmitted at Point(s) of Receipt.

Delivery—The satisfaction of a futures contract position through the tendering and receipt of the actual commodity.
Delivery modes of Natural Gas and Electric Service:
- **Economy**—Energy provided at a lower price when a supplier has excess available capacity.
- **Firm**—Energy provided to customers on a continuous basis without interruption.
- **Interruptible**—Energy provided to customers with a provision that permits curtailment or cessation of service at the discretion of the distributing company under certain circumstances, as specified in the service contract.
- **Off-Peak**—Energy provided to customers in accordance with contractual arrangements to interrupt consumer load at times of peak load by direct control of the utility system operator or by action of the consumer at the direct request of the system operator.
- **Seasonal**—Energy provided to customers in accordance with contractual arrangements to interrupt consumer load at times of seasonal (i.e., summer and winter) peak load by direct control of the utility system operator or by action of the consumer at the direct request of the system operator.

Delivery Month—The month specified in a given futures contract for delivery of the actual spot or cash commodity.

Delivery Notice—A notice presented through an exchange clearinghouse by a clearing member advising of the intention to deliver the actual commodity in satisfaction of contract obligations.

Delivery Point—See Point(s) of Delivery

Delta—The sensitivity of an option's value to a change in the price of the underlying futures contract, also referred to as an option's futures-equivalent position. Deltas are positive for bullish options positions, or calls, and negative for bearish options positions, or puts. Deltas of deep in-the-money options are approximately equal to one; deltas of at-the-money options are 0.5; and deltas of deep out-of-the-money options approach zero.

Delta Neutral Spread—A spread where the total delta position on the long side and the total delta position on the short side add up to approximately zero. Also known as neutral spread.

Demand—The rate at which electric energy is delivered to or by a system or part of a system, generally expressed in kilowatts or megawatts, at a given instant or averaged over any designated interval of time. Demand should not be confused with Load. Types of Demand include:

- **Instantaneous Demand**—The rate of energy delivered at a given instant.
- **Average Demand**—The electric energy delivered over any interval of time as determined by dividing the total energy by the length of time in the interval (e.g., 20 kWh used in ¼ hr has average demand = $20/\frac{1}{4}$ = 80 kW).
- **Integrated Demand**—The average of the instantaneous demands over the demand interval.
- **Interruptible Demand**—The magnitude of customer demand that, in accordance with contractual arrangements, can be interrupted by direct control of the system operator or by action of the customer at the direct request of the system operator. In some instances, the demand reduction may be initiated by the direct action of the system operator (remote tripping) with or without notice to the customer in accordance with contractual provisions. Interruptible Demand as defined here does not include Direct Control Load Management.
- **Demand Interval**—The time period during which electric energy is measured, usually in 15-, 30-, or 60-minute increments.
- **Peak Demand**—The highest electric requirement occurring in a given period (e.g., an hour, a day, month, season, or year). For an electric system, it is equal to the sum of the metered net outputs of all generators within a system and the metered line flows into the system, less the metered line flows out of the system.

- **Coincident Demand**—The sum of two or more demands that occur in the same demand interval.
- **Non-coincident Demand**—The sum of two or more demands that occur in different demand intervals.
- **Contract Demand**—The amount of capacity that a supplier agrees to make available for delivery to a particular entity and which the entity agrees to purchase.
- **Firm Demand**—That portion of the Contract Demand that a power supplier is obligated to provide except when system reliability is threatened or during emergency conditions.
- **Billing Demand**—The demand upon which customer billing is based as specified in a rate schedule or contract. It may be based on the contract year, a contract minimum, or a previous maximum and, therefore, does not necessarily coincide with the actual measured demand of the billing period.

Demand bid—A bid into the power exchange indicating a quantity of energy or an ancillary service that an eligible customer is willing to purchase and, if relevant, the maximum price that the customer is willing to pay.

Demand Charge—A fee based on the peak speed (i.e., kW) at which electricity used during the billing cycle. Residential customers are generally not levied a demand charge.

Demand Destruction—Reductions in demand that involve more than momentary responses to pricing; e.g., drop in demand for natural gas by fertilizer plants when they are shut down due to sustained high gas pricing.

Demand Response—Ability of end user to cut back on power use when called on to do so by an RTO or Load Serving Entity.

Demand-side Management (DSM)—Refers to the planning, implementation, and monitoring of activities designed to encourage consumers to modify patterns of electricity usage, including the timing and level of electricity demand. Covers the complete range of load-shape objectives, including strategic conservation and load management, as well as strategic load growth.

Demand-side Management Costs—When applied to a utility, they are the costs incurred to achieve the capacity and energy savings from a Demand-Side Management Program. When applied to a customer, they may include all costs to control demand, especially peak demand.

Demonstrated Maximum Net Capacity (DMNC)—Maximum net generation output of a generation source, averaged over an ISO-defined period.

Demonstration—The application and integration of a new product or service into an existing or new system. Most commonly, demonstration involves the construction and operation of a new electric technology interconnected with the electric utility system to demonstrate how it interacts with the system. This includes the impacts the technology may have on the system and the impacts that the larger utility system might have on the functioning of the technology.

Demurrage—Compensation paid for detention of a ship during loading or unloading beyond the scheduled time of departure.

Department of Energy (DOE)—The federal government agency engaged in establishing policies and programs relating to national energy matters.

Derating (Generator)—A reduction in a generating unit's Net Dependable Capacity. Types of derating include:
>**Forced Derating**—An unplanned component failure (immediate, delayed, postponed) or other condition that requires the output of the unit be reduced immediately or before the next weekend.
>**Maintenance Derating**—The removal of a component for scheduled repairs that can be deferred beyond the end of the next weekend, but requires a reduction of capacity before the next planned outage.
>**Planned Derating**—The removal of a component for repairs that is scheduled well in advance and has a predetermined duration.
>**Scheduled Derating**—A combination of maintenance and planned deratings.

Deregulation—The elimination of regulation from a previously regulated industry or sector of an industry.

Derivatives—A specialized security or contract that has no intrinsic overall value, but whose value is derived from an underlying security (e.g., stock, bond, currency, or commodity) or factor (such as an index). A generic term that, in the energy field, may include options, futures, swaps, forwards, etc.

DG—See Distributed Generation

Diesel fuel—A fuel composed of distillates obtained in petroleum refining operation or blends of such distillates with residual oil used in backup, emergency, and peaking power plants.

Differential—The amount/degree of a difference, or a change in an amount.

Differentials—Price differences between classes, grades and locations of different stocks of the same commodity. In futures, the discounts (or premiums) allowed for grades or locations of a commodity lower (or higher) than the par or a basis grade or location specified in the futures contract; also called allowances.

Direct Access—The ability of a retail customer to purchase commodity electricity directly from the wholesale market rather than through a local distribution utility. See also Retail Competition.

Direct Control Load Management—The customer demand that can be interrupted by direct control of the system operator controlling the electric supply to individual appliances or equipment on customer premises. This type of control, when used by utilities, usually involves residential customers. Direct Control Load Management as defined here does not include Interruptible Demand. Also called automated demand response (ADR).

Direct Customer (DC)—(e.g. large industrial) that buys power directly from an ISO-managed wholesale market.

Direct Load Control—Refers to program activities that can interrupt consumer load at the time of annual peak load by direct control of the utility system operator by interrupting power supply to individual appliances or equipment on consumer premises. Direct Load Control excludes Interruptible Load and Other Load Management effects. One of the most common control methods involves the cycling or temporary shutdown of electric domestic hot water heaters through automatic remote control.

Disaggregated Virtual Trading (DVT)—occurs when one buys both day-ahead and real-time power so as to hedge the real-time (i.e., hour-ahead) against the day-ahead. It is akin to taking opposing positions in a commodity so one is covered regardless of which way that commodity's price moves. One goal of fostering this process at the wholesale level is to encourage convergence of the day-ahead market (DAM) and hour-ahead market (HAM) so that there is less difference between the volatility of the two (in the NY ISO, the HAM is significantly more volatile than the DAM).

Disaggregation—The functional separation of the vertically integrated utility into smaller, individually owned business units (i.e., generation, dispatch/control, transmission, distribution). The terms "deintegration," "disintegration" and "delamination" are sometimes used to mean the same thing. See also Divestiture. As applied to a customer's loads, the ability to perceive separate load profiles for each major load.

Disco—See Distribution Utility

Discounting—Recalculation of the value of a trade or contract based on the time value of money or other factors.

Discretionary Account—An arrangement by which the holder of an account gives written power of attorney to someone else, often a broker, to buy or sell without prior approval of the account holder.

Dismissal—Action taken by a regulatory commission to end a proceeding through an order. If a proceeding is dismissed without prejudice, the applicant may make a revised filing for the case. If the proceeding is dismissed with prejudice, the applicant must file a request for rehearing.

Dispatch/Economic Redispatch—Choosing and/or deploying generators based on their cost of operation or posted market price for power

Dispatchable Generation—Generation available physically or contractually to respond to changes in system demand or to respond to transmission security constraints. See Must-Run Generation.

Distillate Fuel Oil—A general classification for one of the petroleum fractions produced in conventional distillation operations. It is used primarily for space heating, on-and-off-highway diesel engine fuel (including railroad engine fuel and fuel for agriculture machinery), and electric power generation. Included are Fuel Oils No. 1, No. 2, and No. 4; and Diesel Fuels No. 1, No. 2, and No. 4. No. 2 is the most commonly consumed distillate fuel.

Distributed Generation (DG)—A system utilizing small generators located on a utility's distribution system for the purpose of meeting local (substation level) peak loads and/or displacing the need to build additional (or upgrade) local distribution lines. With the advent of small-scale generation (i.e., less than several MW), distributed generation includes customer and/or marketer-owned capacity feeding single loads (e.g., a chiller plant) customers, or small groups of customers.

Distribution—The delivery of electricity to the retail customer's home or business through low voltage distribution lines (typically in the 120 volt to 69,000 volt range).

Distribution provider (electric)—Provides and operates the wires between the transmission system and the end-use customer. For those end-use customers who are served at transmission voltages, the Transmission Owner also serves as the Distribution Provider.

Distribution System—The portion of an electric system that is dedicated to delivering electric energy directly to an end user.

Distribution System Operator (DSO)—A utility managing its distribution system in a manner that allows (and possibly encourages) development and operation of distributed energy resources (DER) on that system. Acting much like an ISO does on a transmission system, it may also manage hourly energy pricing at the distribution level to achieve goals set by a Public Utility Commission (PUC). This concept is a key to New York State's "Reforming the Energy Vision" (REV) proposals initiated in 2015.

Distribution Utility (Disco)—The regulated electric utility entity that constructs and maintains the distribution wires connecting the transmission grid to the final customer. The Disco can also perform other services such as aggregating customers, purchasing power supply and transmission services for customers, billing customers and reimbursing suppliers, and offering other regulated or non-regulated energy services to retail customers. The "wires" and "customer service" functions provided by a distribution utility could be split so that two totally separate entities are used to provide these two types of distribution services.

Disturbance—An unplanned event that produces an abnormal system condition.

Diversity Exchange—An exchange of capacity or energy, or both, between systems whose peak loads occur at different times.

Diversity Factor—The ratio of the sum of the coincident maximum demands of two or more loads to their non-coincident maximum demand for the same period.

Divestiture—The stripping off of one utility function from the others by selling or in some other way changing the ownership of the assets related to that function. Most commonly associated with selling generation assets so they are no longer owned by the shareholders that own the transmission and distribution assets. See also Disaggregation.

Dividend Yield—Percentage derived by dividing the latest indicated annual dividend by the closing price of the security, and then multiplying by 100.

DMNC—See Demonstrated Maximum Net Capacity

Docket—A formal proceeding before the Commission. Docket numbers are assigned to individual proceedings.

DOE—See Department of Energy

Doji—A commodity trading term indicating that the daily opening and closing price of a commodity (such as oil) were the same.

Drafting—Release of water through turbines other than those used to produce hydropower.

DSM—See Demand-side Management

DSR (Demand-side Resources)—See EDRP (Emergency Demand Response Program)

Dual-fuel—Ability to use more than one fuel to operate a boiler, chiller, generator, or other energy-related equipment.

DVT—See Disaggregated Virtual Trading

Dynamic Rating—The process that allows a system element rating to vary with the changing environmental conditions in which the element is located.

Dynamic Schedule—A telemetered reading or value that is updated in real time and used as a schedule in the Automatic Generation Control/Area Control Error equation and the integrated value of which is treated as a schedule. Commonly used for "scheduling" commonly owned generation or remote load to or from another Control Area.

Dynamic Scheduling Service—See Ancillary Services

EBI—Electronic billing information.

EBIT—Earnings before interest and taxes.

EBITDA—Earnings before interest, taxes, depreciation, and amortization. It is a standard way to measure a company's operating performance, irrespective of its tax environment and financial decision-making.

ECEMP—See Embedded Costs Exceeding Market Prices

Economic Development Zone (EDZ)—An area designated by a government agency in which businesses financial incentives are available, such as reduced taxes. Electric and natural gas utilities may provide discounts to customers in these zones.

Economic Dispatch—The allocation of demand to individual generating units on line to effect the most economical production of electricity.

Economic Efficiency—A term that refers to the optimal production and consumption of goods and services. This generally occurs when prices of products and services reflect their marginal costs. Economic efficiency gains can be achieved through cost reduction, but it is better to think of the concept as actions that promote an increase in overall net value (which includes, but is not limited to, cost reductions).

Economies of Scale—Advantages realized when combining of loads or services results in decreased average long-run costs.

EDI—See Electronic Data Interchange

Edison Electric Institute (EEI)—An association of electric companies formed in 1933 "to exchange information on industry developments and to act as an advocate for utilities on subjects of national interest."

EDL—see Environmental Disclosure Label

EDRP—See Emergency Demand Response Program

EDZ—See Economic Development Zone

EEI—See Edison Electric Institute

Effective Date—The date on which a rate schedule or tariff sheet becomes legally effective

EFP—See Exchange of Futures for Physicals

EFS—See Exchange of Futures for Swaps

EFT—See Electronic Fund Transfer

EIM—See Energy Imbalance Market

ELCON—See Electricity Consumers Resources Council

Electric industry restructuring—The process of replacing a monopolistic system of electric utility suppliers with competing sellers, allowing individual retail customers to choose their supplier but still receive delivery over the power lines of the local utility. It includes the reconfiguration of vertically integrated electric utilities.

Electric Plant (Physical)—A facility containing prime movers, electric generators, and auxiliary equipment for converting mechanical, chemical, and/or fission energy into electric energy.

Electric Power Association—General term for an organization representing a group of co-op utilities before state and federal regulatory boards; may also be part of the name of a co-op (e.g., Valley Electric Power Association)

Electric Power Supply—Electricity required to meet the customers' needs, including energy, operating capacity, losses, ancillary services, and installed capacity including installed reserves.

Electric Rate Schedule—Also often called a tariff, a statement of the electric rates and the terms and conditions governing its application, including attendant contract terms and conditions that have been accepted by a regulatory body with appropriate oversight authority. See also tariff.

Electric System Losses—Total electric energy losses in the electric system. The losses consist of transmission, transformation, and distribution losses between supply sources and delivery points. Electric energy is lost primarily due to resistance heating of transmission and distribution elements.

Electric Utilities—All enterprises engaged in the production and/or distribution of electricity for use by the public, including investor-owned electric utility companies; cooperatively owned electric utilities; government-owned electric utilities (municipal systems, federal agencies, state projects, and public power districts).

Electric Utility—A corporation, person, agency, authority, or other legal entity or instrumentality that owns or operates facilities for the generation, transmission, distribution, or sale of electric energy primarily for use by retail customers and is defined as a utility under the statutes and rules by which it is regulated. Types of Electric Utilities include investor-owned, cooperatively owned, and government-owned (federal agency, crown corporation, state, provincials, municipals, and public power districts). This term includes the Tennessee Valley Authority, but does not include other Federal power marketing agencies.

Electrical Energy—The generation or use of electric power by a device over a period of time, expressed in kilowatt-hours (kWh), megawatt hours (MWh), or gigawatt hours (GWh). Some ways of buying/supplying energy are:
- Economy Energy—Electrical Energy produced and supplied from a more economical source in one system and substituted for that being produced or capable of being produced by a less economical source in another system.
- Emergency Energy—Electrical Energy purchased by a member system whenever an event on that system causes insufficient Operating Capability to cover its own demand requirement.
- Firm Energy—Electrical Energy backed by capacity, interruptible only on conditions as agreed upon by contract, system reliability constraints, or emergency conditions and where the supporting reserve is supplied by the seller.
- Non-firm Energy—Electrical Energy that may be interrupted

by either the provider or the receiver of the energy by giving advance notice to the other party to the transaction. This advance notice period is equal to or greater than the minimum period agreed to in the contract. Non-firm Energy may also be interrupted to maintain system reliability of third-party Transmission Providers. Non-firm Energy must be backed up by reserves.

- Off-Peak Energy—Electrical Energy supplied during a period of relatively low system demands as specified by the supplier.
- On-Peak Energy—Electrical Energy supplied during a period of relatively high system demands as specified by the supplier.

Electricity—A form of energy characterized by the presence and motion of elementary charged particles generated by friction, induction, or chemical change.

Electricity broker—An entity that arranges the sale and purchase of electric energy, the transmission of electricity, and/or other related services between buyers and sellers but does not take title to any of the power sold.

Electricity congestion—A condition that occurs when insufficient transmission capacity is available to implement all of the desired transactions simultaneously.

Electricity Consumers Resources Council (ELCON)—An association of 28 large industrial consumers of electricity accounting for over five percent of all electricity consumed in the United States. Formed in 1976 "to enable member companies to work cooperatively for the development of coordinated, rational and consistent policies affecting electric energy supply and pricing at the federal, state, and local levels."

Electricity demand—The rate at which energy is delivered to loads and scheduling points by generation, transmission, and distribution facilities.

Electricity demand bid—A bid into the power exchange indicating a quantity of energy or an ancillary service that an eligible customer is willing to purchase and, if relevant, the maximum price that the customer is willing to pay.

Electricity generation—The process of producing electric energy or the amount of electric energy produced by transforming other forms of energy, commonly expressed in kilowatt-hours (kWh) or mega-watt-hours (MWh).

Electricity sales—The amount of kilowatt-hours sold in a given period of time; usually grouped by classes of service, such as residential, commercial, industrial, and other. "Other" sales include sales for public street and highway lighting and other sales to public au-thorities, sales to railroads and railways, and interdepartmental sales.

Electronic Data Interchange (EDI)—Process for communicating billing data in one of several standard electronic formats used in a cus-tomer's Accounts Payable software.

Electronic Fund Transfer (EFT)—Process for shifting funds between a bank and a supplier or customer via secure internet-based com-munication.

Element—Any electric device with terminals that may be connected to other electric devices, such as a generator, transformer, circuit, circuit breaker, or bus section

Embedded Costs Exceeding Market Prices (ECEMP)—Embedded costs of utility investments exceeding market prices are: 1) costs incurred pursuant to a regulatory or contractual obligation; 2) costs that are reflected in cost-based rates; and 3) cost-based rates that exceed the price of alternatives in the marketplace. ECEMPS may become "stranded costs" where they exceed the amount that can be recovered through the asset's sale or employment in a com-petitive marketplace. Regulatory questions involve whether such costs should be recovered by utility shareholders and if so, how they should be recovered. "Transition costs" are stranded costs which are charged to utility customers through some type of fee or surcharge after the assets are sold or separated from the vertically integrated utility. "Stranded assets" are assets which cannot be sold for some reason.

Emergency—Any abnormal system condition that requires automatic or immediate manual action to prevent or limit loss of transmission facilities or generation supply that could adversely affect the reliability of the electric system.

Emergency Demand Response Program (EDRP)—Any of several efforts to utilize nontraditional resources, such as retail customer-owned generation and curtailable power demands (e.g., dimming systems) to increase the reserve margin between peak demand and peak available capacity; such programs are typically operated through ISOs via retail power providers, such as utilities and retail power suppliers.

Eminent Domain—A legal process used as a last resort if a landowner and the project proponent cannot reach agreement on compensation for use or purchase of property required for the project. The project proponent is still required to compensate the landowner for the use or purchase of the property, and for any damages incurred during construction. However, the level of compensation would be determined by a court according to state law.

End User—A retail customer of a natural gas or electricity product or service.

End-use Services—The provision of energy, power and related services, such as energy efficiency or on-site generation, to the ultimate consumer.

Energy Charge—That portion of the charge for electric service based upon the electric energy (kWh) consumed or billed.

Energy Deliveries—Energy generated by one electric utility system and delivered to another system through one or more transmission lines.

Energy Efficiency—The act of using less energy/electricity to perform a given function, as distinguished from conservation, which implies accepting less.

Energy Emergency—A condition when a system or power pool does not have adequate energy resources (including water for hydro units) to provide its customers' expected energy requirement. See Capacity Emergency.

Energy Intensity—A ratio of energy consumption to another metric, typically national gross domestic product in the case of a country's energy intensity. Sector-specific intensities may refer to energy consumption per household, per unit of commercial floor space, per dollar value industrial shipment, or another metric indicative of a sector. Improvements in energy intensity include energy efficiency and conservation as well as structural factors not related to technology or behavior.

Energy Exchange—Transaction whereby the receiver accepts delivery of energy for a supplier's account and returns energy later at times, rates, and in amounts as mutually agreed. See Storage; Banking.

Energy Glossary—Energy terms and definitions used in Energy Information Administration reports and by the energy industry.

Energy Imbalance Market (EIM)—pricing mechanism established to handle transactions for correcting sub-hourly energy imbalances such as occur from renewable sources.

Energy Imbalance Service—Provides energy correction for any hourly mismatch between a transmission customer's energy supply and the demand served.

Energy Policy Act of 1992 (EPACT)—This legislation created a new class of power generators, "exempt wholesale generators," that are exempt from the provisions of the Public Utilities Holding Company Act of 1935 and grants the authority to the Federal Energy Regulatory Commission to order and condition access by eligible parties to the interconnected transmission grid.

Energy Receipts—Energy generated by one electric utility system and received by another system through one or more transmission lines.

Energy Service Company (ESCo)—A company that offers to reduce a client's energy cost, often by taking a share of such reduced costs as repayment for installing and financing such upgrades. In some states, ESCos are also suppliers of energy.

Energy Service Provider—California's name for an ESCo; often abbreviated ESP.

Energy Source—The primary source that provides the power that is converted to electricity through chemical, mechanical, or other means. Energy sources include coal, petroleum and petroleum products, gas, water, uranium, wind, sunlight, and geothermal heat.

Energy—The capacity for doing work. Electrical energy is usually measured in kilowatt-hours (kWh), while heat energy is usually measured in British thermal units (Btu).

Enhanced Oil Recovery (EOR)—Processes utilizing (e.g.) compressed carbon dioxide (CO_2) gas to force oil otherwise not recoverable from played-out wells or seams previously deemed not economically viable. The CO_2 would remain underground, thus making it part of a carbon capture and sequestration (CCS) effort.

Enron Terms—Terms introduced by the Enron controversy.

Environmental Assessment—An Environmental Assessment (EA) evaluates the consequences of a proposed action on the environment and recommends measures to minimize any potentially adverse affects. An EA is prepared when the environmental scoping process has determined that the project would not significantly affect the quality of the environment, as versus an Environmental Impact Statement.

Environmental Disclosure Label (EDL)—A listing of a power supplier's energy sources for its power production, typically by the percent of electricity produced from each. Some states require that EDLs be filed by all utilities and competitive power suppliers each year.

Environmental Impact Statement—A statement required of federal agencies by Section 102 (C) of the National Environmental Policy Act of 1969 for major federal actions that could significantly impact the quality of the environment.

Environmental Protection Agency (EPA)—A federal agency charged with protecting the environment.

EOI—Expression Of Interest; Australian/British term for what is essentially a RFP.

EOR—See Enhanced oil recovery.

EPA—See Environmental Protection Agency

EPACT—See Energy Policy Act of 1992

Equity Capital—The sum of capital from retained earnings and the issuance of stocks.

Escalation—A clause, usually in long-term supply contracts, which provides for periodic price adjustment based on variations in any or all cost factors. "Escalating prices" are the opposite of "firm" prices, which are not subject to change over the life of a contract.

ESCo—See Energy Service Company

Estimated Usage—Customer usage based on history rather than meter readings.

e-Tag—An electronic label developed by NERC to track wholesale power trades across and to control areas. It includes the following information: date and time of the transaction, the physical path taken between the point of generation and the point of consumption (typically a utility, not an end user), the financial path (who sold the power and who bought it), the flow (MWhr in each hour of transfer) and the OASIS transmission requests for each power system that the flow will cross.

European Option—An options contract that may be exercised only on its expiration date. See also American Option.

Evaluative Resolution—Determination of who is right or wrong during dispute resolution

Evergreen Clause—A section in a contract under which the contract is automatically renewed (typically for the same time length as the existing contract) unless the buyer, within a period stated in contract (e.g., at least 60 days before the end of the existing term), notifies the seller that he does not wish the contract renewed. Failure by the seller to notify the buyer that the "evergreen period" has started or is approaching does not typically release the buyer from the automatic renewal, unless so stated in the contract.

EWG—Exempt Wholesale Generator

Exchange Member—An individual or organization holding a seat on an exchange. See seat.

Exchange of Futures for Physicals (EFP)—A futures contract provision involving the delivery of physical product (which does not necessarily conform to contract specifications in all terms) to one market participant from another and a concomitant assumption of equal and opposite futures positions by the same participants to the physical transaction. An EFP occurs during the futures contract trading period.

Exchange of Futures for Swaps (EFS)—A bilateral negotiated transaction where a physically delivered commodity futures contract is exchanged for a swap position in the same or a related commodity. Such deals are common when dealing with non-standard energy contracts not otherwise available in regulated markets.

Exempt Wholesale Generator (EWG)—Created under the 1992 Energy Policy Act, these wholesale generators are exempt from certain financial and legal restrictions stipulated in the Public Utilities Holding Company Act (PUHCA) of 1935.

Exercise—The process of converting an options contract into a futures position.

Exercise Price—The price at which the underlying futures contract will be bought or sold in the event an option is exercised. Also called the strike price.

Exotic—A complex or unusual contractual arrangement; opposite of Vanilla.

Expected Unserved Energy—The expected amount of energy curtailment per year due to demand exceeding available capacity. It is usually expressed in megawatt hours (MWh).

Expenditure—The incurrence of a liability to obtain an asset or service.

Expiration Date—In options trading, the last day on which an option can be exercised or traded; if not exercised, it expires without having yielded any financial value.

Expression of Interest (EOI)—Australian/British term for what is essentially a RFP.

Extrinsic Value—The amount by which the premium exceeds its intrinsic value. Also known as time value.

Facility—An existing or planned location or site at which prime movers, electric generators, and/or equipment for converting or using mechanical, chemical, and/or nuclear energy into electric energy are situated, or will be situated. A facility may contain more than one generator of either the same or different prime mover type. For a cogenerator, the facility includes the industrial or commercial process.

FASB—See Financial Accounting Standards Board (this abbreviation is pronounced "faz-bee")

Fat Boy—An Enron term for taking advantage of differential market pricing. When adjacent power markets exhibit significantly differing prices (such as the price-controlled California market and the nearby Pacific Northwest where no price caps existed), a trader would purchase power at a low (controlled) price and sell it into the uncapped market for several times that price because there was no limit on such interstate transactions or price differentials (until FERC mandated regional price caps).

Fault—An event occurring on an electric system such as a short circuit, a broken wire, or an intermittent connection.

FCR—Abbreviation for financial congestion rights. See also transmission congestion contracts (TCC) and fixed transmission rights (FTR).

Federal Energy Regulatory Commission (FERC)—The federal agency regulating price, terms and conditions of power sold in interstate commerce and that of all transmission services. FERC is the federal counterpart to state utility regulatory commissions (often called PUCs).

Federal Power Act—Administered by FERC, it was enacted in 1920 and amended in 1935. It incorporated the Federal Water Power Act administered by the former Federal Power Commission, whose activities were confined almost entirely to licensing non-Federal hydroelectric projects. Other parts extended the Act's jurisdiction to include regulating the interstate transmission of electrical energy and rates for its sale as wholesale in interstate commerce.

Federal Power Commission (The predecessor agency to FERC)— The Federal Power Commission (FPC) was created by an Act of Congress under the Federal Water Power Act on June 10, 1920. Originally charged with regulating the electric power and natural gas industries, it was abolished on September 20, 1977, when the Department of Energy was created. The functions of the FPC were then divided between the newly created Department of Energy and the Federal Energy Regulatory Commission.

Federal Rates—Rates that apply to the marketing of wholesale power and transmission services provided by government owned or leased facilities to non-Federal customers. This is done through FERC-approved rate schedules or contracts at revenue levels sufficient to repay the Federal charges incurred in providing these services

Federally Mandated Congestion Charges (FMCC)—Line item seen on some electric bills that covers increased costs due to locational marginal pricing (as required by many ISOs, which are federally chartered organizations); seen first in Connecticut in early 2004.

Feebates—A revenue neutral strategy which imposes a fee on polluting resources and rebates those fees to cleaner technologies.

Feed-in tariff—A pricing mechanism that pays certain distributed generation systems (such as PV) a much higher price for the power they produce and "feed in" to the distribution grid. The higher price provides an immediate and proportional incentive based on actual power production. Funding comes from ratepayer revenue.

Fence—A long (short) underlying position together with a long (short) out-of-the-money put and a short (long) out-of-the-money call. All options must expire at the same time.

FERC—See Federal Energy Regulatory Commission

FERC Order 745—A generic proceeding for setting pricing for demand response (DR) resources, e.g., how much to pay DR providers based on the value resulting from a reduction in grid demand. The Order generally set that number to be equivalent to the value of wholesale hourly price of power paid to wholesale power providers, as charged at the time the DR resources are utilized.

FERC Order 888—A 1996 generic proceeding that requires open access to transmission facilities to address undue discrimination and to bring more efficient, lower cost power to the nation's electricity consumers.

FERC Order 1000—A 2011 generic proceeding requiring transmission planning at the regional level that considers and evaluates possible transmission alternatives to produce a regional transmission plan. It requires fair allocation to beneficiaries of costs of transmission solutions chosen to meet regional transmission needs.

FIA—Futures Industry Association.

Fill—The price at which an order is executed (applies to any kind of market).

Financial Accounting Standards Board—Federal agency that creates standards for reporting expenses and profits that impact both commodity and energy services accounting. Often abbreviated as FASB, pronounced "faz-bee."

Financial Congestion Rights (FCR)—New England ISO term for Transmission Congestion Contract, also known as Congestion Revenue Rights (CRRs)

Fire Wall—The line of demarcation separating residential and small commercial customers from all other customers preventing the shifting of costs between customer classes as a result of special rate discounts or other restructuring activities.

Firm Gas—Gas sold on a continuous and generally long-term contract.

Firm Power—Power or power producing capacity intended to be available at all times during the period covered by a guaranteed commitment to deliver, even under adverse conditions.

Firm Purchase—A purchase of electricity or natural gas by one utility from another under contract, arranged in advance of delivery.

Firm Transmission Rights (FTR)—PJM Interconnection (ISO) term for transmission congestion contract; also called fixed (or financial) transmission rights (FTR) or financial congestion rights (FCR).

Firm Transmission Service—Point-to-point transmission service that is reserved and/or scheduled for a term of one year or more and that is of the same priority as that of the Transmission Provider's firm use of the transmission system. Firm Transmission service that is reserved and/or scheduled for a term of less than one year shall be considered Short-Term Firm Transmission Service for the purposes of service liability.

First Notice Day—The first day on which a notice of intent to deliver the actual commodity or security can be given to the clearing house by the party that sold the contract.

Fish Flush—Description of increased movement of water at hydroelectric power plants in which the flow of water may bypass generators to allow fish to move without harm.

Fixed Cost—A set cost, under terms of contract, even if no energy was produced.

Flexible ramping—An ISO-level process for covering costs of generators whose output deviates from their hourly scheduled output, such as wind turbines. The extra costs to accommodate such variations (e.g., maintaining spinning reserve from natural gas generators).

Flexible rate—An economic incentive rate designed by a Public Utility Commission that allows a utility to negotiate costs with industrial or large commercial users.

Floor broker—An exchange member who executes orders for the accounts of others.

Floor trader or local—An exchange member who executes orders for his own account.

Flowgate Model—Proposed (2000) alternative method to price transmission in which various points of constrained transmission would be assigned economic value such that the sum of the cost to transmit power passing through a combination of "flow gates" would become the charge added to the commodity cost for power to bring it to an LDC's distribution system. This model may compete with the existing Locational Marginal Pricing (LMP) method.

Flue Gas Desulfurization Unit (Scrubber)—Equipment used to remove sulfur oxides from the combustion gases of a boiler plant before discharge to the atmosphere. Chemicals, such as lime, are used as the scrubbing media.

Flue-gas particulate collector—Equipment used to remove fly ash from the combustion gases of a boiler plant before discharge to the atmosphere. Particulate collectors include electrostatic precipitators, mechanical collectors (cyclones), fabric filters (baghouses), and wet scrubbers.

Fly ash—Particulate matter mainly from coal ash in which the particle diameter is less than 1×10^{-4} meter (.0001 meters). This ash is removed from the flue gas using flue gas particulate collectors such as fabric filters and electrostatic precipitators.

FMCC—Federally Mandated Congestion Charges

FOB (free on board)—A transaction in which the seller provides a product at an agreed unit price, at a specified loading port within a specified period; it is the responsibility of the buyer to arrange for transportation and insurance and lift the material within a specified time period.

Force majeure—A standard clause which indemnifies either or both parties to a transaction whenever events reasonably beyond the control of either or both parties occur to prevent fulfillment of the terms of the contract.

Forced outage—The shutdown of a generating unit, transmission line, or other facility for emergency reasons or a condition in which the generating equipment is unavailable for load due to unanticipated break-down.

Forecast—Predicted demand for electric power. A forecast may be short term (e.g., 15 minutes) for system operation purposes, long-term (e.g., five to 20 years) for generation planning purposes, or for any range in between.

Forecast Uncertainty—Probable deviations from the expected values of factors considered in a forecast.

Forex—Foreign exchange, usually referring to currency trading.

Form Value Trading—See Tolling.

Forward Curve—Commodity pricing for future months that may, if purchased now, be secured for fixed price contracts with terms covering those months. Such curves may be developed from pricing seen at private and public exchanges, marketer quotes, and other information, gathered and portrayed in wholesale trade publications (e.g., *Megawatt Daily*).

Forward Price Curve—A set of commodity prices typically based on the market prices for futures available at a given point in time. (Same as forward curve.)

Forward Resource Auction (FRA)—MISO term for process to secure sufficient capacity to meet demand in parts of MISO where retail access is available (e.g., lower Illinois).

Forwards—A forward is a commodity bought and sold for delivery at some specific time in the future. It is differentiated from futures markets by the fact that a forward contract is customized, non-exchange traded, and a non-regulated hedging mechanism.

Fossil Fuel—Any naturally occurring organic fuel, such as petroleum, coal, and natural gas.

Fossil-Fuel Plant—A plant using coal, petroleum, or gas as its source of energy.

FPA (Federal Power Act of 1935)—Established guidelines for federal regulation of interstate energy sales. It is the primary statute governing FERC regulation of the electric sector.

FPC—See Federal Power Commission, precursor to FERC.

Frequency—Rate of variation of voltage and current, typically 60.0 Hertz (i.e., cycles per second), except during a time correction.

Frequency Bias—A value, usually given in megawatts per 0.1 Hertz (MW/0.1 Hz), associated with a Control Area that relates the difference between scheduled and actual frequency to the amount of generation required to correct the difference.

Frequency Deviation—A departure from scheduled frequency.

Frequency Error—The difference between actual system frequency and the scheduled system frequency.

Frequency Regulation—The ability of a Control Area to assist the interconnected system in maintaining scheduled frequency. This assistance can include both turbine governor response and automatic generation control.

Frequency Response—The ability of a power supplier to respond to an undesired change in the alternating current waveform on a grid. To be viable, that ability must be automatic and able to maintain power adjustments within 5 to 30 seconds of the frequency fluctuation. That response is calculated by summing the change in demand and the change in generation, and dividing it by the change in frequency, expressed in megawatts per 0.1 Hertz (MW/0.1 Hz).

FTR—See Firm Transmission Rights

FUCO—Foreign utility company owning power generation in the U.S.

Fuel—Any substance that can be burned to produce heat; also, materials that can be fissioned in a chain reaction to produce heat.

Fuel Adjustment Charge—Fee added to base utility rates to account for above-normal pricing of fuels used in the supply of electricity, natural gas, or district steam.

Fuel Diversity—The situation in which a given supply portfolio is made up of plants using several different types of fuel to generate electricity, for the purpose of avoiding over-reliance on one fuel, and the related risk of supply interruption and price spikes. A form of hedging by maintaining a diverse portfolio of fuel inputs.

Fuel Expenses—These costs include the fuel used in the production of steam or driving another prime mover for the generation of electricity. Other associated expenses include unloading the shipped fuel and all handling of the fuel up to the point where it enters the first bunker, hopper, bucket, tank, or holder in the boiler house structure.

Full-Forced Outage—The net capability of main generating units that are unavailable for load for emergency reasons.

Functional Unbundling—The functional separation of generation, transmission, and distribution transactions within a vertically integrated utility without selling or spinning off these functions into separate companies. See also divestiture.

Fundamentals analysis—The study of pertinent supply and demand factors which influence the specific price behavior of commodities.

Fungible—Interchangeable. Products which can be commingled for purposes of shipment or storage. Also means an ability to easily exchange for money or other assets.

Futures Contract—A standardized financial agreement for the purchase or sale of a commodity or product at a specified price that is traded in an open auction under the rules of an exchange and that requires delivery on or settlement through the sale or purchase of an offsetting contract by a specified future date.

Futures Market—Arrangement through a contract for the delivery of a commodity at a future time, place, and price specified at the time of purchase. The price is based on an auction or market basis. Futures are a standardized, exchange-traded, and government-regulated hedging tool.

Gamma—The sensitivity of an option's delta to changes in the price of the underlying futures contract.

Gas—Typically natural gas (which consists primarily of methane); a fuel burned by boilers and by internal combustion engines for electric generation. May also include manufactured, waste, and coal gas.

Gas oil—European designation for No. 2 heating oil and diesel fuel.

Gas Turbine Plant—A power plant in which the prime mover is a gas turbine. A gas turbine consists typically of an axial-flow air compressor, one or more combustion chambers, where liquid or gaseous fuel is burned and the hot gases are passed to the turbine and where the hot gases expand to drive the generator and are then used to run the compressor.

Gasification—A process that converts liquefied natural gas (LNG) from a liquid back to a gas. This is done by increasing the temperature and decreasing the pressure of the LNG. May also mean a process for processing coal to produce combustible gas.

Gathering Line—Pipelines, generally small in diameter, used to transport oil or gas from the well to a processing facility or a mainline pipeline.

Generating Unit—Any combination of physically connected generator(s), reactor(s), boiler(s), combustion turbine(s), or other prime mover(s) operated together to produce electric power.

Generation (Electricity)—The process of producing electrical energy from other forms of energy; also, the amount of electric energy produced, usually expressed in kilowatt-hours (kWh) or megawatt hours (MWh).

Generation Company (Genco)—A regulated or non-regulated entity (depending upon the industry structure) that operates and maintains existing generating plants. The Genco may own the generation plants or interact with the short term market on behalf of plant owners. In the context of restructuring the market for electricity, Genco is sometimes used to describe a specialized marketer for the generating plants formerly owned by a vertically integrated utility.

Generation Dispatch and Control—Aggregating and dispatching (i.e., directing to specific locations and/or loads) generation from various generating facilities, providing backup and reliability services.

Generation, Gross—The electrical output at the terminals of the generator, usually expressed in megawatts (MW).

Generation, Net—Gross generation minus station service or unit service power requirements, usually expressed in megawatts (MW).

Generator—A machine that converts mechanical energy into electrical energy.

Generator capacity—The maximum output, commonly expressed in megawatts (MW), that generating equipment can supply to system load, adjusted for ambient conditions.

Generator Nameplate Capacity—The full-load continuous rating of a generator, prime mover, or other electric power production equipment under specific conditions as designated by the manufacturer. Installed generator nameplate rating is usually indicated on a nameplate physically attached to the generator.

Gentailer—A wholesale generating company that also owns a retail power supply company. Such entities are able to create and sell electricity to an end use customer without being regulated as a utility.

Geothermal energy—Hot water or steam extracted from geothermal reservoirs in the earth's crust. Water or steam extracted from geothermal reservoirs can be used for geothermal heat pumps, water heating, or electricity generation.

Geothermal Plant—A plant using heat generated from within the earth in which the prime mover is a steam turbine. The turbine is driven either by steam produced from hot water or by natural steam found at various depths beneath the earth's surface. The energy is extracted by drilling into pressurized areas and/or pumping fluids that transfer such heat to the turbine.

Get Shorty—An Enron term for manipulating pricing in ancillary service markets. Ancillary Services are required to move power and consist of such actions as running a generator at low speed even when it is not supplying power to the grid in order to have the ability to rapidly ramp up to meet a short term need. Such a service is called spinning reserve. The ISO must contract for such services to maintain balance between supply and demand, and separate submarkets exist to trade such services. The Get Shorty tactic involved agreeing to provide ancillary services in the state's day-ahead market at a high price but doing so by buying those services from others at a lower price in the hour-ahead market. The tactic involved providing false information regarding the actual source (e.g., generator) of the actual ancillary service supply.

Gigajoule (GJ)—One billion joules, approximately equal to 948,211 British thermal units, which is approximately equivalent to the energy in 1 decatherm of natural gas.

Gigawatt (GW)—One billion watts. Same as 1000 megawatts or 1,000,000 kilowatts.

Gigawatt-hour (GWh)—One billion watt-hours.

GJ (Gigajoule)—See Gigajoule

GLD—See Guaranteed Load Drop

GMC—See Grid Management Charge

Good Till Canceled—An order to be held by a broker until it can be filled or until canceled.

Granularity—The level of detail under which data may be examined or analyzed. A shorter time interval, for example, would be more "granular" than a longer one.

Green Button standard—A voluntary protocol, developed by the US DOE in 2011, to portray quarter-hourly electric interval data. Adopted by many utilities, software firms, etc., it fosters use and interchange of such data for load analysis and control.

Green Markets/Marketing—Sales and purchases of power from renewable or otherwise environmentally desirable resources and efficiency services.

Greenhouse Effect—The increasing mean global surface temperature of the earth caused by gases in the atmosphere (including carbon dioxide, methane, nitrous oxide, ozone, and chlorofluorocarbon). The greenhouse effect allows solar radiation to penetrate but absorbs the infrared radiation instead of returning it to space.

Grid—A system of interconnected power lines and generators that is managed so that the generators are dispatched as needed to meet the requirements of the customers connected to the grid at various points.

Grid Management Charge (GMC)—Fee charged by an ISO to owners of on-site generation (such as cogeneration plants) as a standby charge for possible use of the transmission system if such a plant goes down and must take power from the grid.

Grid-supply tariff—A replacement for net metering, it allows grid-connected sites to export excess (typically renewable) power produced on-site to the grid and receive financial credits (less than that previously provided by net metering).

Gridco—An independent company responsible for the operation of a grid; now often called a transco.

Gross Generation—The total amount of electric energy produced by the generating units at a generating station or stations, measured at the generator terminals. See also Net Generation.

Guaranteed Load Drop (GLD)—Measures the MW reduction (for DR purposes) that occurred from the customer's baseline load, which is a hypothetical estimate of how much electricity the customer would have consumed had it not followed a demand response dispatch. This may also apply to an aggregate group of customers by combining the MW reductions of the group. The GLD ensures that all of a demand response provider's curtailments are counted in the capacity market, even if the ISO had only intended the customer to reduce from a defined MW amount to a specified lower MW amount.

GW (Gigawatt)—See Gigawatt

GWh (Gigawatt-hour)—See Gigawatt-hour

HAM—See Hour Ahead Market

Hammer, The—High speed trading software used to execute "banging the close," a technique designed to influence the closing price of derivatives contracts for fuels on the Globex part of the NY Mercantile Exchange in 2008 and later. To do so, the program's algorithm would gather some orders to be placed just before the closing bell, while holding the others to be placed right at the closing point. The CFTC caught the firm doing this and, 4 years later, fined it $15 million.

HCA—See Host Control Area

Heat content—The amount of heat energy available to be released by the transformation or use of a specified physical unit of an energy form (e.g., a ton of coal, a barrel of oil, a kilowatt-hour of electricity, a cubic foot of natural gas, or a pound of steam). The amount of heat energy is commonly expressed in British thermal units (Btu). Note: Heat content of combustible energy forms can be expressed in terms of either gross heat content (higher or upper heating value) or net heat content (lower heating value), depending upon whether or not the available heat energy includes or excludes the energy used to vaporize water (contained in the original energy form or created during the combustion process). The Energy Information Administration typically uses gross heat content values.

Hedge Bucket—Term coined in early 2000 to describe accounting procedure that matches investments in financial instruments (e.g., futures, options) with the specific gain or loss resulting from that investment, as required under FASB Statement 133.

Hedge ratio—1) Ratio of the value of futures contracts purchased or sold to the value of the cash commodity being hedged, a computation necessary to minimize basis risk. 2) The ratio, determined by an options delta, of futures to options required to establish a riskless position. For example, if a $1/dekatherm change in the underlying gas futures price leads to a \$0.25/dthm change in the option premium, the hedge ratio is 4 (four options contracts for each futures contract).

Hedging—The simultaneous initiation of equal and opposite positions in the cash and futures markets. Hedging is employed as a form of financial protection against adverse price movement in the cash market.

Hedging Contracts—Contracts which establish future prices and quantities of electricity independent of the short-term market. Derivatives may be used for this purpose. See Contracts for Differences, Forwards, Futures Market, and Options.

Herfindahl-Hirschmann Index (HHI)—Widely used measure of market concentration calculated by summing the squares of the percentages of a market as represented by each competitor. For example, 5 equally sized generators each contributing 20% of a market would result in that market having an HHI of 2000 (5×20^2 = 2000). Ten equally sized generators each contributing 10% of a market would have an HHI of 1000 (i.e., $10 \times 10^2 = 1000$). Federal agencies handling anti-trust issues consider a market having an HHI of 1800 or more to be highly concentrated. While useful when examining competition among generators, it is of limited use when dealing with interacting components (such as generation and transmission) where even a low HHI would not necessarily reveal monopolistic capabilities.

HHI—See Herfindahl-Hirschmann Index

HID—A form of lighting (high-intensity discharge) based on high voltage charging of various gases: typically used for outdoor and industrial illumination.

High—The highest price of the day of trading for a particular stock or commodity

Highly controlled—Coal-fired generation that is not "highly controlled" lacks both sulfur dioxide scrubbers and devices to control the emission of nitrogen oxide.

Hockey Stick Pricing (or Bidding)—In hourly power markets, pricing is paid to the level needed to meet hourly demand. This may create an exponential price curve that, to many, looks like a hockey stick; i.e., a slightly sloping line followed by a suddenly steep curve. Some suppliers bid low during high demand periods to ensure they are selected, but are then paid based on the high price taken from the highest accepted bidder, or (knowing there's a good chance that their capacity will be taken) bid very high at such times, thus setting the market clearing price, and making a bundle in the process..

Horizontal spread—Calendar or time spread.

Host Control Area (HCA)—A control area that confirms and implements scheduled interchange for a transmission customer that operates generation or serves customers directly within the control area's metered boundaries. The control area within whose metered boundaries a commonly owned unit or terminal is physically located.

Hour Ahead Market (HAM)—Wholesale power pricing provided in 1-hour blocks; effectively the same thing as real time pricing (RTP).

Hourly Metering—A type of interval metering where the measurement or recording of customer usage is collected in 60-minute intervals. The competitive metering model is based upon the implementation of hourly metering of customers or the application of standard rate class load profiles.

Hourly Non-Firm Transmission Service—Point-to-point transmission that is scheduled and paid for on an as-available basis and is subject to interruption.

Hub—Point at which gas or electricity may be re-routed into major interstate pipelines or transmission lines; pricing for either commodity is typically based at a hub, with transportation to the customer's utility being an additional charge.

Huff-and-puff—an enhanced oil recovery process involving injection of carbon dioxide (CO_2) into an oil well to increase output. Also seen (by the US DOE in 2010) as a way to assess carbon sequestration potential for a geologic formation.

Hydroelectric Plant—A plant in which the turbine generators are driven by falling or rushing water.

Hydrogen—The lightest of all gases, occurring chiefly in combination with oxygen in water; exists also in acids, bases, alcohols, petroleum, and other hydrocarbons.

Hydropower—Converting the energy of running water into electricity.

Hydropower Exemption—Small hydropower projects (less than 5 megawatts) that are exempted from licensing requirements. Exemptions are granted in perpetuity.

Hydropower License (Major)—A major license authorizing the licensee to construct and operate a new project, or continue to operate an existing project, which is capable of generating more than 5 megawatts. The license term is from 30 to 50 years.

Hydropower License (Minor)—A minor license issued by the Commission, authorizing the licensee to construct and operate a new project, operate an existing project, which is capable of generating less than 5 megawatts. The license term is typically 30 years.

Hydropower Re-license—A major or minor license that authorizes the licensee to receive a new license term for an existing licensed project. The new license term may be from 30 to 50 years.

IBT—See Internal Bilateral Transaction

ICAP—See Installed Capacity, under Capacity

ICAP Deficiency Charge—Also called simply the "ICAP charge," this fee is levied by some ISOs on retail power marketers unable to serve their load with their own generation capacity. Such marketers may need to secure capacity from the ISO to ensure its committed availability to meet contracted load. If, for example, a marketer owned no generation and was serving a 100 MW load, he would need to either contract for capacity under a bilateral contract with a generator or buy that 100 MW of capacity from the ISO at its defined ICAP deficiency charge. This charge is determined in ways that differ among ISOs. Some portray the charge as a means to encourage construction of new capacity, but the success of that process is a matter of contention.

ICAP Tag—Reservation of generation capacity needed to serve a specified load (also known in some ISOs as a "capacity and transmission" tag).

IDR—See Interval Data Recorder

Imbalance—A condition where the generation and interchange schedules do not match demand.

Implied heat rate—A calculation of the day-ahead electric price divided by the day-ahead natural gas price. Implied heat rate is also known as the 'break-even natural gas market heat rate,' because only a natural gas generator with an operating heat rate (measure of unit efficiency) below the implied heat rate value can make money by burning natural gas to generate power. Natural gas plants with a higher operating heat rate cannot make money at the prevailing electricity and natural gas prices.

Inadvertent Energy Balancing—A Control Area's accounting of its inadvertent interchange, which is the accumulated difference between actual and scheduled interchange.

Inc-ing—Enron term for "increasing" prices when demand in a day-ahead spot market (operated by an ISO or power exchange) was underestimated by utilities and power traders scheduling load. When demand is underestimated, the day-ahead wholesale price of power is artificially driven down. When an ISO finds that demand is actually higher than it predicted based on faulty schedules submitted by some marketers (or by extreme weather or plant outages), the ISO must purchase enough extra power to maintain the balance between supply and demand. Since it must do so on short notice, marketers can then bid up short-term power pricing, forcing the ISO to pay a higher price to maintain that balance.

Incremental Energy Cost—The additional cost that would be incurred by producing or purchasing the next available unit of electrical energy above the current base cost.

Incremental Heat Rate—The amount of additional heat that must be added to a thermal generating unit at a given loading to produce an additional unit of output. It is usually expressed in British thermal units per kilowatt hour (Btu/kWh) of output.

Independent Market Monitor (IMM)—an entity (typically a technical consulting firm) serving an ISO or public utility commission to watch a power market for pricing abuses and other problems; essentially the same as a Market Monitoring Unit (MMU).

Independent Power Producer (IPP)—A private entity that operates a generation facility and sells power to electric utilities for resale to retail customers. As used in NERC reference documents and reports, any entity that owns or operates an electricity generating facility that is not included in an electric utility's rate base. This term includes, but is not limited to, cogenerators and small power producers and all other non-utility electricity producers, such as exempt wholesale generators who sell electricity. Another category of generator used in a generic manner to refer to QFs and EWGs. Some IPPs are excluded from PUHCA constraints because they are structured so that no one person owns 10 percent or more of the generator and thus no owner is classified as a holding company.

Independent System Operator (ISO)—A neutral, independent, federally regulated and (typically non-profit) organization with no financial interest in generating facilities that administers the operation and use of the transmission system. ISOs exercise final authority over the dispatch of generation to preserve reliability and facilitate efficiency, ensure non-discriminatory access, administer transmission tariffs, ensure the availability of ancillary services, and provide information about the status of the transmission system and available transmission capacity. Under some proposals, an ISO may make some transmission investment decisions. See also Control Area.

Industrial Sector—Generally defined as manufacturing, construction, mining, agriculture, fishing, and forestry establishments (Standard Industrial Classification [SIC] codes 01-39). A utility may classify industrial service using the SIC codes, or based on demand or annual usage exceeding some specified limit. The limit may be set by the utility based on the rate schedule of the utility. See also Standard Industrial Classification.

Integrated merchants—Power suppliers that include companies owning both unregulated and regulated distribution companies. Also known as hybrids.

Integrated Resource Planning (IRP)—A public planning process administered by regulatory agencies within which costs and benefits of both demand- and supply-side resources are evaluated to develop the least-total-cost mix of utility resource options. In many states, IRP includes a means for considering environmental damages caused by electricity supply/transmission and identifying cost-effective energy efficiency and renewable energy alternatives. IRP has become a formal process prescribed by law in some states and under some provisions of the Clean Air Act Amendments of 1992.

Integrated Resource Planning Principles—The underlying principles of IRP can be distinguished from the formal process of developing an approved utility resource plan for utility investments in supply- and demand-side resources. A primary principle is to provide a framework for comparing a variety of supply- and demand-side and transmission resource costs and attributes outside of the basic provision (or reduction) of electric capacity and energy. These resources may be owned or constructed by any entity and may be acquired through contracts as well as through direct investments. Another principle is the incorporation of risk and uncertainty into the planning analysis. The public participation aspects of IRP allow public and regulatory involvement in the planning rather than the siting stage of project development.

Interchange—Electric power or energy that flows from one entity to another. Some types of interchange are:

Actual Interchange—Metered electric power that flows from one entity to another.

Interchange Scheduling—The actions taken by scheduling entities to arrange transfer of electric power. The schedule consists of an agreement on the amount, start and end times, ramp rate, and degree of firmness.

Scheduled Interchange—Electric power scheduled to flow between entities, usually the net of all sales, purchases, and wheeling transactions between those areas at a given time.

Interconnected Operations Services (IOS)—Services that transmission providers may offer voluntarily to a transmission customer under Federal Energy Regulatory Commission Order No. 888 in addition to Ancillary Services. See also Ancillary Services.

Backup Supply Service—Provides capacity and energy to a transmission customer, as needed, to replace the loss of its generation sources and to cover that portion of demand that exceeds the generation supply for more than a short time.

Dynamic Scheduling Service—Provides the metering, telemetering, computer software, hardware, communications, engineering, and administration required to electronically move a transmission customer's generation or demand out of the Control Area to which it is physically connected and into a different Control Area.

Real Power Loss Service—Compensates for losses incurred by the Host Control Area(s) as a result of the interchange transaction for a transmission customer. Federal Energy Regulatory Commission's Order No. 888 requires that the transmission customer's service agreement with the Transmission Provider identify the entity responsible for supplying real power loss.

Restoration Service—Provides an offsite source of power to enable a Host Control Area to restore its system and a transmission customer to start its generating units or restore service to its customers if local power is not available.

Interconnected System—A system consisting of two or more individual electric systems that normally operate in synchronism and have connecting tie lines.

Interconnection—When their names are capitalized, any one of the five major electric system networks in North America: Eastern, Western, ERCOT, Quebec, and Alaska. When not capitalized, the facilities that connect two systems or Control Areas. Additionally, an interconnection refers to the facilities that connect a non-utility generator to a Control Area or system.

Interdepartmental Service (Electric)—Interdepartmental service includes amounts charged by the electric department at tariff or other specified rates for electricity supplied by it to other utility departments.

Interface—The specific set of transmission elements between two areas or between two areas comprising one or more electrical systems.

Intermediary Control Area—A Control Area that has connecting facilities in the scheduling path between the sending and receiving Control Areas and has operating agreements that establish the conditions for the use of such facilities.

Intermediate Load (Electric System)—The range from base load to a point between base load and peak. This point may be the mid-point, a percent of the peak load, or the load over a specified time period.

Internal Bilateral Transaction (IBT)—The purchase or sale of electric energy or regulation obligations between two market participants internal to an ISO/market.

Internal Combustion Plant—A power plant in which the prime mover is an internal combustion engine. An internal combustion engine has one or more cylinders in which the process of combustion takes place, converting energy released from the rapid burning of a fuel-air mixture into mechanical energy. Diesel or gas-fired engines are the principal types used in electric plants. The plant is usually operated during periods of high demand for electricity.

Interruptability—The right of a utility to interrupt power delivery to a customer based on an existing contract, tariff, or agreement, typically during a system emergency or when wholesale market prices exceed a defined level.

Interruptible Gas—Gas sold to customers with a provision that permits curtailment or cessation of service at the discretion of the distributing company under certain circumstances, as specified in the service contract.

Interruptible Load—Refers to program activities that, in accordance with contractual arrangements, can interrupt consumer load at times of seasonal peak load by direct control of the utility system operator or by action of the consumer at the direct request of the system operator. In some instances the load reduction may be affected by direct action of the system operator (remote tripping) after notice to the consumer in accordance with contractual provisions.

Interruptible load or interruptible demand (electric)—Demand that the end-use customer makes available to its Load-Serving Entity via contract or agreement for curtailment.

Interstate—Sales where transportation of natural gas, oil, or electricity crosses state boundaries. Interstate sales are subject to Commission jurisdiction.

Interstate Commerce—An interchange of goods or commodities which involves transportation between states.

Interval data—kWh consumption in short (e.g., 15-minute) intervals, or average kW demands in such intervals.

Interval Data Recorder (IDR)—a typical 'smart' meter or a data logger capable of capturing and recording interval data.

Interval Meter—A meter that measures usage in time increments shorter than those used in billing (e.g., a month). Often synonymous with "hourly meter," "hourly interval meter," or "time-differentiated meter."

Interval Metering—Interval metering is the measurement of customer energy usage by fixed time periods or intervals. Typically, the interval time period is 15 minutes, but can vary according to the customer, tariff and/or T&D system needs. Interval metering is provided to commercial and industrial customers and some residential customers.

Intervenor—An intervenor formally participates in a regulatory proceeding by filing a request to intervene. Intervenors are able to file briefs, appear at hearings, and be heard by the courts if they choose to appeal the regulatory body's final ruling.

In-the-Money—An option that can be exercised and immediately closed out against the underlying market for a cash credit. The option is in-the-money if the underlying futures price is above a call option s strike price, or below a put option s strike price.

Intra-Control Area Transaction—A transaction from one or more generating sources to one or more delivery points where all the sources and delivery points are entirely within the metered boundaries of the same Control Area.

Intrastate—Sales where transportation of natural gas, oil, or electricity occur within a single state and do not cross state boundaries. Intrastate sales are not subject to Commission jurisdiction.

Intrinsic Value—The amount by which an option is in-the-money. An option which is not in-the-money has no intrinsic value. For calls, intrinsic value equals the difference between the underlying futures price and the option's strike price. For puts, intrinsic value equals the option s strike price minus the underlying futures price. Intrinsic value is never less than zero.

Inverted Market—A futures market in which nearby months are trading at a premium to distant months. See also Backwardation.

Investor-owned Utility (IOU)—A company, owned by stockholders for profit, that provides utility services. A designation used to differentiate a utility owned and operated for the benefit of shareholders from municipally owned and operated utilities and rural electric cooperatives.

Invitation to Bid (ITB)—Much like a Request for Proposals (RFP) except that the only item generally requested is a price, not a statement of services involving negotiation.

IOU—See Investor-owned Utility

IPP—See Independent Power Producer

IRP—See Integrated Resource Planning

ISDN (Integrated Services Digital Network)—A 128 Kbps (kilobytes per second) digital telephone service initially used to speed up Internet access. Largely abandoned for that purpose by the late 1990s when DSL and fiber optic cable TV service vastly improved access speeds and allowed web-based digital TV transmission.

Island—A portion of a power system or several power systems that is electrically separated from the interconnection due to the disconnection of transmission system elements.

Islanding—Problem occurring when a distributed generation (DG) power source remains on-line during a utility grid failure, thereby energizing local distribution lines and creating a safety hazard. Also used to mean the ability of a facility to power itself independent of the grid.

ISO—See Independent System Operator

ITB—See Invitation to Bid

ITP—Independent Transmission Provider

IVR—Interactive Voice Response; phone-based system for automatically responding to queries via speech recognition software.

JACE—Java Application Control Engine; embedded software (pioneered and promoted by Iridium Inc.) that allows a variety of devices and protocols to be controlled via a common distributed automation system.

Japanese candlestick chart—See Candlestick Chart

Jet Fuel—Kerosene/gasoline mixture; quality kerosene product used primarily as fuel for commercial turbojet and turboprop aircraft engines and single-cycle turbine generators, e.g., JP-4 fuel.

JOA—See Joint Operating Agreement

Joint Action Agency—A group of municipal utilities that may act as a supplier of generation and transmission capacity to serve its members needs. In essence, it may be a company owned by its members that acts as a wholesale aggregator of their loads in order to secure the lowest cost services for them through shared purchase and ownership of power plants and transmission lines.

Joint Operating Agreement (JOA)—Contract between two ISOs to co-ordinate border and other common issues related to transmission, dispatching, etc.

Joint Unit Control—Automatic generation control of a generating unit by two or more entities.

Joule—metric unit of energy; 1055 joules = 1 Btu; a gigajoule (GJ) is roughly the same as 1 million Btu (i.e., about 948,000 Btu), and thus about the same as a dekatherm.

Jurisdictional—Utilities, ratepayers and regulators (and impacts on those parties) that are subject to state regulation in a state considering restructuring.

Kerosene—A light petroleum distillate that may be used in electric peaking plants or as an alternate fuel in jet engine turbines used in power production.

Key Performance Indicator (KPI)—Business term for a criterion that relates to a desired outcome; e.g., a KPI for energy efficiency might be a Portfolio Manager score or a threshold energy use per square foot per year.

Khazzoom-Brookes Postulate—Theory (put forth in 1979-80 as an argument against appliance efficiency standards) stating that more efficient consumption of energy reduces the unit cost of goods, leading to increased consumption, which then leads to new applications, which eventually enter that lower price range. This "rebound effect" results in more—not less—energy being consumed. In a nutshell, "instead of capping energy demand, what we observe is that improvements in energy efficiency lead to ever and ever-greater levels of energy usage." Taking this idea to its extreme would result in a consumer buying and driving a second car or appliance, not because it is needed, but merely because the first one was more fuel-efficient. It should also be noted that, simultaneous to introduction of efficiency measures, new regulations and codes (e.g., ASHRAE) on how energy is utilized may limit and/or cap how intensely energy gets used in a new appliance or device, thus blocking excessive use merely because it has become cheaper.

Kilowatt (kW)—One thousand watts; a unit of energy usage rate or transfer rate.

Kilowatt-hour (kWh)—The basic unit of electric energy equal to one kilowatt of power supplied to, or taken from, an electric circuit steadily, for one hour. Also known as consumption. When fully converted to heat, equivalent to 3,412 Btu.

Kirchhoff's Law—A description regarding how power moves among multiple electric paths (such as transmission lines) based on their equivalent resistance, regardless of any contractual or other theoretical arrangement. Such movement may result in parallel loop flows that can hinder or complicate the transfer of power from one location to another.

kVar (kilovar)—The unit of measure of reactive power—the power supplied to most types of electromagnetic equipment, such as motors. This unit is often used when charging a large power customer for low power factor.

kW—See kilowatt

kWh—See kilowatt-hour

Lambda—Also called "system lambda," a term describing the cost of the next kilowatt hour that could be produced from dispatchable units on an electric supply system.

Last Notice Day—The final day on which notices of intent to deliver on futures contracts may be issued.

Last Trading Day—The final day on which futures contracts may be traded, after which they must be settled by delivery.

Laytime—How long an item remains at a physical point before it must be moved to market or to a customer.

LBMP—Locational-Based Market Pricing. See Locational Market Pricing

LCOE—See Levelized Cost of Electricity

LDA—See Locational Deliverability Area

LDC—See Local Distribution Company

Levelized Cost Of Electricity (LCOE)—Total cost in $/kWh including fuel, capital, maintenance, and interest on capital amortized across the expected lifetime of the generating source. LCOE is a standard utility metric for comparing the price of new sources of power when determining which should be added as needed. Also called "levelized cost of energy."

Leverage Ratio—A measure that indicates the financial ability to meet debt service requirements and increase the value of the investment to the stockholders (i.e., the ratio of total debt to total assets).

LFG—Landfill gas; an impure form of natural gas commonly taken from landfills to power onsite generation (Btu content may range from 30%-65% that of utility-grade natural gas).

Liability—An amount payable in dollars or by future services to be rendered.

License Plate Rate—A concept popularized by ELCON in 2000 in which transmission rates within an RTO would remain the same as a transmission customer presently pays to a utility regardless of where the power moves within the RTO. This differs from pancake rates in which, to move across several utilities, the customer would pay each separate utility's rate as his power crossed that utility's territory.

Lignite—The lowest rank of coal, often referred to as brown coal, used almost exclusively as fuel for steam-electric power generation. It is brownish-black and has a high inherent moisture content, sometimes as high as 45 percent. The heat content of lignite ranges from 9 to 17 million Btu per ton on a moist, mineral-matter-free basis. The heat content of lignite consumed in the United States averages 13 million Btu per ton, on the as-received basis (i.e. containing both inherent moisture and mineral matter).

Limit—The maximum amount a futures price may advance or decline in any one day's trading session.

Limit Order—A contingent order for a futures trade specifying a certain maximum (or minimum) price, beyond which the order is not to be executed.

Line Congestion—See Congestion

Line Loss—typically a percent quantifying energy lost during delivery; also called Unaccounted For Energy (UFE), that may include losses due to theft of service.

Liquefied Natural Gas (LNG)—Reducing the temperature of natural gas to minus 259 degrees at atmospheric pressure will convert the gas into a liquid. Its volume as a liquid is about 1/600 compared to its volume as a gas at standard temperature and pressure.

Liquefied Petroleum Gas (LPG)—Propane, butane or propane-butane mixtures derived from crude oil refining or natural gas fraction-ation.

Liquidation—The closing out of long positions. (More generally, liquidation means the conversion of something into cash or another commodity.)

Liquidity—How easily assets (such as forward contracts) may be converted to cash. A futures market is said to be "liquid" when it has a high level of trading activity allowing buying and selling of futures contracts with minimum price disturbance.

LMP—See Locational Marginal Pricing

LNG—See Liquefied Natural Gas

Load (Electric)—The amount of electric power delivered or required at any specific point or points on a system, as demanded by an electricity-consuming system, or systems.

Load Building—Refers to programs that are aimed at increasing the usage of existing electric equipment or the addition of electric equipment. Examples include industrial technologies such as induction heating and melting, direct arc furnaces and infrared drying; cooking for commercial establishments; and heat pumps for residences.

Load Centers—A geographical area where large amounts of power are drawn by end users.

Load Cycle—The normal pattern of demand over a specified time period associated with a device or circuit.

Load Data—Electric power consumption (kWh) as measured in one-hour intervals.

Load Duration Curve—A non-chronological, graphical summary of demand levels with corresponding time durations using a curve, which plots demand magnitude (power) on one axis and percent of time that the magnitude occurs on the other axis at or above that magnitude. Often called a histogram.

Load Factor—A measure of the degree of uniformity of demand over a period of time, equivalent to the ratio of average demand to peak demand expressed as a percentage. It is calculated by dividing the total energy provided by a system during the period by the product of the peak demand during the period and the number of hours in the period.

Load Following—An electric system's process of regulating its generation to follow the changes in its customers' demand.

Load Forecast—An estimate of the level of future energy needs. A "bottom-up" forecast uses utility revenue meters to develop system-wide loads; used often in projecting loads of specific customer classes. "Top-down" uses utility meters at generation and transmission sites to develop aggregate control area loads; useful in determining reliability planning requirements, especially where retail choice programs are not in effect.

Load Management—Utility activities designed to influence the timing and amount of electricity that customers may use.

Load Pocket—An area that is served by local generators when the existing electric system is not able to provide full service, typically due to transmission constraints.

Load Profile—Electric power consumption (kWh) as measured in one-hour (or shorter) intervals. Also called "load shape."

Load Serving Entity (LSE)—See Local Distribution Company.

Load Shedding—The process of deliberately removing (either manually or automatically) preselected customer demand from a power system in response to an abnormal condition to maintain the integrity of the system and minimize overall customer outages.

Load Shift—An Enron term for taking advantage of apparent transmission line congestion. Under Load Shift, a trader over-schedules load to create the appearance of line congestion. He then shifts that load from the congested zone to a less congested zone (by restating his load schedules), thus earning payments from the ISO for relieving the fictitious congestion.

Load Shifting—Demand-side management programs designed to encourage consumers to move their use of electricity from on-peak times to off-peak times.

Local Distribution Company (LDC)—Any firm, other than a natural gas pipeline, engaged in the transportation or local distribution of natural gas and its sale to customers that will consume the gas. Also called Utility Distribution Company (UDC), or Load Serving Entity (LSE).

Locational Deliverability Area (LDA)—PJM term for a region bounded by its transmission constraints on the delivery of capacity into a sub-region. A capacity auction may be held to develop the price of capacity in each LDA, thus ensuring each sub-region has sufficient capacity to meet its projected peak load.

Locational Marginal Pricing (LMP)—Pricing process that raises all wholesale power prices in a zone to a defined level, based on the cost of the most expensive source of power used to serve that zone at a given time. Also called Location Based Marginal Pricing (LBMP).

Locational power exchange—A simultaneous pair of purchase and sales transactions involving the same quantity of power and the same parties, but at two different locations and at two different prices, typically done to avoid FERC regulation of a trade.

Locked market—A market (an exchange or a private group of traders) where prices have reached their daily trading limit and trading may only be conducted at that price.

LOLE—See Loss of Load Expectation

Long—As in to "take a long position," or to be "long on supply," indicates a supplier has commodity to sell; opposite of short, which indicates the supplier has only a minimal amount or no commodity at risk and/or to sell.

Long Ton—An avoirdupois weight measure equaling 2,240 pounds (1,016 kg), not to be confused with metric ton (= 1,000 kg = 2,204 pounds).

Long-range Planning—The process of forecasting long-term loads, determining a reasonable set of potential resources to meet such loads (including reduction of loads through energy efficiency), analyzing the costs (sometimes including externality costs) of several possible mixes of such resources, and identifying the resources to be secured to meet such future needs.

Loop Flows—See Parallel Path Flows

Loss of Load Expectation (LOLE)—The expected number of days in the year when the daily peak demand exceeds the available generating capacity. It is obtained by calculating the probability of daily peak demand exceeding the available capacity for each day and adding these probabilities for all the days in the year. The index is referred to as Hourly Loss-of-Load-Expectation if hourly demands are used in the calculations instead of daily peak demands.

Losses—See Electric System Losses

Loss-of-load-Probability—See Expected Unserved Energy

Low—The lowest price of the day of trading for a particular stock or commodity.

LPG—See Liquefied Petroleum Gas

LSE—See Load Serving Entity and Local Distribution Company

M&A—Mergers and acquisitions; business term describing the consolidations that occur within an industry as participants aggregate with each other, or buy them.

M&V—See Monitoring & Verification

MACT—See Maximum Achievable Control Technology

Manufactured gas—A gas obtained by destructive distillation of coal or by the thermal decomposition of oil, or by the reaction of steam passing through a bed of heated coal or coke. Examples are coal gases, coke oven gases, producer gas, blast furnace gas, blue (water) gas, carbureted water gas. Btu content varies widely.

MAPE—Mean Absolute Percent Error; the standard accuracy forecasting measure when projecting hourly grid loads.

Margin—In electric service, the difference between net capacity resources and net internal demand. Margin is usually expressed in megawatts (MW). In futures trading, funds posted during the trading life of a futures contract to guarantee fulfillment of contract obligations.

Margin Call—The demand for additional (variation) margin, the result of adverse price movement and consequent erosion of equity, usually at an exchange, which is the only entity with the power to make and enforce such a call, wherein a broker who bought a commodity without yet having paid for it is then required to pay a defined amount (usually a percentage) of its price.

Marginal Cost—In the utility context, the cost to the utility of providing the next (marginal) kilowatt-hour of electricity, irrespective of sunk costs.

Marginal Line Loss—actual power lost while sending electricity from one point to another along a transmission line.

Marginal Loss Surplus Allocation (MLSA)—surplus payments made to a power trader due to transmission line loss charges; some market manipulation has occurred wherein traders inflate up-to-congestion trades to maximize MLSA payments to reimburse them for line losses that may increase due to the congestion they helped create.

Market-based pricing—Prices of electric power or other forms of energy determined in an open market system of supply and demand under which prices are set solely by agreement as to what buyers will pay and sellers will accept. Such prices could recover less or more than full costs, depending upon what the buyers and sellers see as their relevant opportunities and risks.

Market clearing price—The price at which supply equals demand for the Day-ahead or hour-ahead markets.

Market order—See At the Market

Market Power—The ability of any market trader with a large market share to significantly control or affect price by withholding production from the market, limiting service availability, or reducing purchases.

Market Rate Option (MRO)—a plan for purchasing or generating power at a least expensive cost to consumers.

Market-based Price—A price set by the mutual decisions of many buyers and sellers in a competitive market.

Market-driven Reliability—Efforts made to maintain or improve reliability as a means to increase or hold onto market share.

Marketer—An entity that has the authority to take title to electrical power generated by itself or another entity and re-sell that power at market-based rates. Also, an agent for generation projects who markets power on behalf of the generator. The marketer may also arrange transmission, firming or other ancillary services as needed. Though a marketer may perform many of the same functions as a broker, the difference is that a marketer represents the generator while a broker acts as a middleman.

Mark-to-Market—Pricing mechanism based on valuing a commodity or contract at its immediate resale value.

Market Resource Alternatives (MRA)—Non-transmission alternatives (e.g., power storage batteries, cogeneration or renewable power facilities) that may be called upon to support grid power, and thus receive financial incentives for that service.

MATS—Mercury and Air Toxics Standards; EPA term for regulations designed to greatly cut emissions of mercury and about a dozen other toxic emissions that come primarily from coal and oil-fired power plants.

Maximum Achievable Control Technology (MACT)—EPA term for state-of-the-art emissions control technology, most relevant to coal-fired utilities and industrial boilers.

Maximum Demand—The greatest of all demands of the load that has occurred within a specified period of time.

Maximum Price Fluctuation—The limit, as set by the rules of a commodity exchange, of the fluctuation in the price of a futures contract during any one trading session.

MCF—A MCF is equivalent to 1,000 cubic feet (of gas).

MCP—Market clearing price, same as MCPE ("market clearing price for electricity).

MDR—Market-driven reliability

MDSP—See Meter Data Service Providers

Meter Data Service Providers (MDSP)—In the competitive metering model, an MDSP is an entity that on behalf of a customer collects, translates and/or analyzes meter data for billing and related purposes.

Mean Reversion—Tendency of pricing to return to an average level.

Measurement Terms—Describes units used to measure energy.

Megawatt (MW)—One million watts.

Megawatt hour (MWH)—One million watt-hours.

Megawatt Laundering—See Ricochet

Member System—An eligible customer operating as a part of a lawful combination, partnership, association, or joint action agency composed exclusively of eligible customers.

Merchant Generator—A power supplier that secures revenue based on sales on an open power market rather than regulated rate-of-return pricing. Ownership may be by a utility outside the service territory, or an independent power producer (IPP). All risk for its development, installation, and operation is borne by its owners, and not by customers.

Meter Data Management Agent—Often abbreviated MDMA, the California name for an agency that receives and handles metering data for power suppliers, the ISO, and customers, and has been certified by the California PUC.

Meter Service Providers (MSP)—MSPs are entities that physically handle meters for purposes such as installation, maintenance, setting and upgrading internal parameters and removal.

Metered Value—A measured electrical quantity that may be observed through telemetering, supervisory control and data acquisition (SCADA), or other means.

Metering—The methods of applying devices that measure and register the amount and direction of electrical quantities with respect to time.

Metering Facility—A facility that measures and registers the amount and direction of natural gas or electricity that flows through the facility

Metric Ton—A weight measure equal to 1,000 kilograms, 2,204.62 pounds, and 0.9842 long tons. For approximate conversion purposes, there are about 7.46 barrels of No. 2 distillate fuel in one metric ton; 8.51 barrels of gasoline and 6.7 barrels of residual fuel.

Microgrid—a group of interconnected loads and distributed energy resources (DER) with clearly defined electrical and geographic boundaries that acts as a single controllable entity with respect to the grid and can connect and disconnect from the grid to enable it to operate in both grid connected or island mode.

Midcontinent Independent System Operator (MISO)—Originally known as the Midwest ISO until Manitoba (Canada) was added, this transmission operator covers parts of 15 central U.S. states from the west side of the Great Lakes to Montana, and in some states down to the Gulf of Mexico.

Mileage—With regard to large energy storage systems (e.g., batteries), the number of MW of energy the storage system provided or absorbed.

Mileage multiplier—The number of MWhrs expected from 1 MW of regulation capacity (e.g., large storage batteries).

Mill—one tenth of a cent (i.e., one thousandth of a dollar) per kWh

Minimum Offer Price Rule (MOPR)—a FERC regulation designed to prevent very low and uncompetitive bids for capacity from entering the market.

Minimum Price Fluctuation—The minimum unit by which the price of a commodity can fluctuate per trade on a commodity exchange.

MMBtu—One million British thermal units; one dekatherm. Approximately equal to 1,000 cubic feet of natural gas.

MMCF—Equivalent to 1,000,000 cubic feet (of gas), or 1,000 Mcf.

MMU—Market monitoring unit; a part of an ISO or public utility commission (or contractor serving such entities) that watches a power market to find and identify pricing abuses and other problems.

Mogas—Industry abbreviation for motor gasoline.

Monitoring & Verification (M&V)—A series of processes for determining the usage and savings resulting from installation and operation of energy efficiency measures.

Monopoly—The only seller with control over market sales.

Monopsony—The only buyer with control over market purchases.

Monte Carlo technique—Mathematical method for assessing risk probabilities.

MOPR—See Minimum Offer Price Rule

Motor Gasoline—A complex mixture of relatively volatile hydrocarbons, with or without small quantities of additives, which have been blended to form a fuel suitable for use in spark-ignition engines.

MOU—Memorandum of Understanding: an abbreviated version of a contractual arrangement, typically entered into prior to development of a full contract to approximately define the points of agreement between two parties.

MRO—See Market Rate Option

MSP—See Meter Service Providers

Muni—See Municipal Utility

Municipal Utility (Muni)—A provider of utility services owned and operated by a municipal government. A muni may or may not own generation, transmission, or distribution facilities.

Municipalization—The process by which a municipal government assumes responsibility for supplying utility service to its constituents. In supplying electricity, the municipality may generate and distribute the power or purchase wholesale power from other generators and distribute it.

Must-run Generation—Generation designated to operate at a specific level and not available for dispatch, generally to maintain voltage, frequency, or supply in areas with constrained transmission. Also known as "reliability must run (RMR)" generation. See Dispatchable Generation.

NERC—See North American Electric Reliability Council

NAERO—See North American Electric Reliability Organization

NAESB—See North American Energy Standards Board

NAICS—See North American Industry Classification System

Naked—A long or short market position taken without having an offsetting short or long position.

Name Give-up Model—During online power trading, this process allows anonymous trading until a purchase or sale is made, at which point the name of the counterparty is revealed (i.e., "given up"). Such counterparties are usually on a list of pre-credit approved firms to avoid trading with parties having questionable credit.

NARUC—See National Association of Regulatory Utility Commissioners

NAS—See Network Access Service

NASUCA—See National Association of State Utility Consumer Advocates

National Association of Regulatory Utility Commissioners (NARUC)—An advisory council composed of governmental agencies of the fifty States, the District of Columbia, Puerto Rico and the Virgin Islands engaged in the regulation of utilities and carriers. "The chief objective is to serve the consumer interest by seeking to improve the quality and effectiveness of public regulation in America."

National Association of State Utility Consumer Advocates (NASUCA)—Includes members from 38 states and the District of Columbia. It was formed "to exchange information and take positions on issues affecting utility rates before federal agencies, Congress and the courts."

National Futures Association (NFA)—Private organization including futures brokers that sets protocols for the certification and handling of personnel selling futures.

National Interest Electric Transmission Corridor (NIETC)—DOE term for a pathway whose importance to the nation transcends local interests.

Native gas—Gas in place at the time that a reservoir was converted to use as an underground storage reservoir in contrast to injected gas volumes.

Native load (electric)—The end-use customers that the Load-Serving Entity is obligated to serve.

Native Load Customers—The wholesale and retail customers on whose behalf the Transmission Provider, by statute, franchise, regulatory requirements, or contract, has undertaken an obligation to construct and operate the Transmission Provider's system to meet the reliable electric needs of such customers.

Natural Gas—A naturally occurring mixture of hydrocarbon and non-hydrocarbon gases found in porous geological formations beneath the earth's surface, often in association with petroleum. The principal constituent is methane.

Natural Gas Act (NGA)—Federal law enacted in 1938 that established FERC's authority to regulate interstate pipelines.

Natural Gas Policy Act (NGPA)—Federal law that updated the NGA and set guidelines for deregulation of new gas supplies and continued regulation of old supplies.

Natural Monopoly—A situation where one firm can produce a given level of output at a lower total cost than can any combination of multiple firms. Natural monopolies occur in industries which exhibit decreasing average long-run costs due to size (economies of scale). According to economic theory, a public monopoly governed by regulation is justified when an industry exhibits natural monopoly characteristics.

Negative pricing—When wholesale market pricing for power is less than $0/MWh, typically during off-peak hours in late night hours. A wind farm, for example, receiving the production tax credit of $22/MWh might post a price of -$5/MWh (in essence paying a wholesale buyer $5 per MWh to take the power off his hands). That wind farm would nevertheless receive $17/MWh (= $22 + -$5) in the process.

Negotiated Rate—An alternative to traditional cost-of-service rates, where a rate for a service varies from a pipeline's otherwise applicable tariff and is mutually agreed upon by a pipeline and its customer. At the time a customer is considering a negotiated rate, a recourse rate that is on file in the pipeline's tariff must also be available to that customer. Also known as negotiated/recourse rates.

NERTO—See North East Regional Transmission Organization

Net Capability—The maximum load carrying ability of the equipment, exclusive of station use, under specified conditions for a given time interval, independent of the characteristics of the load. (Capability is determined by design characteristics, physical conditions, adequacy of prime mover, energy supply, and operating limitations such as cooling and circulating water supply and temperature, headwater and tailwater elevations, and electrical use.)

Net Capacity Resource—The total owned capacity, plus capacity available from independent power producers, plus the net of total capacity purchases and sales, less the sum of inoperable capacity, and less planned outages.

Net Dependable Capacity—The maximum capacity a unit can sustain over a specified period modified for seasonal limitations and reduced by the capacity required for station service or auxiliaries.

Net Energy for Load—The electrical energy requirements of an electric system, defined as system net generation, plus energy received from others, less energy delivered to others through interchange. It includes system losses but excludes energy required for storage at energy storage facilities.

Net Generation—Gross generation minus plant use from all electric utility-owned plants. The energy required for pumping at a pumped storage plant is regarded as plant use and must be deducted from the gross generation. See also gross generation.

Net Internal Demand—The metered net outputs of all generators within a system, plus the metered line flows into the system, less the metered line flows out of the system, less direct control load management and, less interruptible demand.

Net metering—process through which electricity generated by an on-site system, such as photovoltaic (PV) panels, counts against billing for power taken from a utility when the on-site system at times generates excess power beyond the on-site need, thus feeding back energy into the utility distribution system. This may be accomplished using special meters that measure flow in both directions, or by using two meters: one that measures total kWh taken by the facility and the other measures power generated by the on-site system. The latter is subtracted from the former to derive the net kWh taken from the utility. If the on-site system generates more than the facility uses during a billing period, a payment from the utility for the difference is issued to the customer.

Net metering aggregation—where a customer has meters at two or more sites where on-site power generation systems (e.g., PV panels) exist, the customer is permitted to aggregate the output of those systems to offset the customer's aggregate energy use at those sites. May also be called remote net metering.

Net Position—A position not offset by a countervailing position; i.e., the opposite of a spread.

Net Present Value (NPV)—Calculation of an asset's worth, taking into account how the value of money will change over the lifetime of an asset.

Net Schedule—The algebraic sum of all scheduled transactions across a given transmission path or between Control Areas for a given period or instant in time.

Net Summer Capability—The steady hourly output, which generating equipment is expected to supply to system load exclusive of auxiliary power, as demonstrated by tests at the time of summer peak load.

Net Winter Capability—The steady hourly output which generating equipment is expected to supply to system load exclusive of auxiliary power, as demonstrated by tests at the time of winter peak load.

Network Access Service (NAS)—Description of service level to local distribution companies under which they may transmit power at a fixed price anywhere within a network; sometimes described as "license plate" pricing

Network Customers—Entities receiving transmission service pursuant to the terms of the transmission provider's network integration tariff.

Network Integration Transmission Service (NITS)—Network integration transmission service allows a transmission customer to integrate, plan, economically dispatch, and regulate its network resources to serve its network load in a manner comparable to that in which the transmission provider utilizes its transmission system to serve its native load customers. Network integration transmission service also may be used by the transmission customer to deliver non-firm energy purchases to its network load without additional charge. In PJM, a transmission fee that is additional to capacity for supply pricing. Most utilities charge it as part of distribution, but several (AEP in Ohio) charged it in commodity power pricing until late 2015.

Network Meter Reading (NMR)—NMR is a type of AMR in which a network is used to collect, transmit and analyze meter data. Some advanced meter functions may be implemented in the network rather than in the meter at the service end point. See also AMR (automatic meter reading).

Network Service—Ability to move power anywhere within a given region (typically a NERC region or ISO) without regard to exact origin or end point.

Network Use Charge (NUC)—charge levied mainly on residential and small commercial customers generating power from solar PV panels to cover use of the distribution network, as separate from utility generation charges. Doing so allows the utility to recover costs of its system even when power taken by the customer is very small due to significant PV output.

New Gas—Gas produced from new formations and fields or drilling after April 1977.

New York Control Area (NYCA)—The part of the State whose transmission and generation assets are under control of the NY ISO.

New York Mercantile Exchange (NYMEX)—A physical marketplace in which various commodities (including electricity, oil, and natural gas) are changed.

NFA—National Futures Association

NGA—See Natural Gas Act

NGPA—See Natural Gas Policy Act

NIETC—See National Interest Electric Transmission Corridor

NITS—See Network Integration Transmission Service

NMR—See Network Meter Reading

Nodal Pricing—Wholesale pricing at transmission nodes (e.g., substations) to better reflect and price congestion costs than occurs via zonal pricing.

Nodes—Point at which transmission lines intersect with each other or interface with distribution systems (often substations containing transformers that step down voltage).

Non-bypassable Charge—Any of a number of charges that would apply to all end-users of electricity, and could not be bypassed except by totally disconnecting from the grid. Includes systems benefits charges, public goods charges, wires charges, and access charges. Typically a fee for use of the wires or access to the grid.

Non-coincident Peak Load—The sum of two or more peak loads on individual systems that do not occur in the same time interval. Meaningful only when considering loads within a limited period of time, such as a day, week, month, a heating or cooling season, and usually for not more than one year.

Non-core Customer—A customer who has several fuel choices and does not rely only on a utility as a resource for continued heat and other energy services.

Non-evaluative Resolution—No one is deciding (evaluating the evidence of) a case for the disputants during dispute resolution.

Non-firm Power—Power or power producing capacity supplied or available under a commitment having limited or no assured availability.

Non-firm Transmission Service—Point-to-point transmission service that is reserved and/or scheduled on an as-available basis and is subject to interruption. Non-firm Transmission Service is available on a stand-alone basis as either Hourly Non-firm Transmission Service or Short-Term Non-firm Transmission Service.

Non-jurisdictional—Utilities, ratepayers and regulators (and impacts on those parties) other than the state-regulated utilities, regulators and ratepayers in a jurisdiction considering restructuring. Examples include utilities in adjacent state and non-state regulated, publicly owned utilities within restructuring states.

Non-public Documents—Includes information location about proposed or existing critical energy infrastructure information (CEII), and privileged information, which is usually confidential business information or cultural resource reports.

Non-utility Generator (NUG)—A generation facility owned and operated by an entity who is not defined as a utility in that jurisdictional area.

Non-utility Power Producer—A corporation, person, agency, authority, or other legal entity or instrumentality that owns electric generating capacity and is not an electric utility. Nonutility power producers include qualifying cogenerators, qualifying small power producers, and other nonutility generators (including independent power producers) without a designated franchised service area, and which do not file forms listed in the Code of Federal Regulations, Title 18, Part 141.

NOPR—See Notice of Proposed Rulemaking

Normal Market—A market is deemed "normal" when carrying charges are reflected in higher prices for increasingly distant futures contract months.

North American Electric Reliability Council (NERC)—A council formed in 1968 by the electric utility industry to promote the reliability and adequacy of bulk power supply in the electric utility systems of North America. NERC consists of eleven regional reliability councils and encompasses essentially all the power regions of the contiguous United States, Canada, and Mexico. The NERC Regions are: ASCC—Alaskan System Coordination Council ECAR—East Central Area Reliability Coordination Agreement ERCOT—Electric Reliability Council of Texas MAIN—Mid-America Interconnected Network MAAC—Mid-Atlantic Area Council MAPP—Mid-Continent Area Power Pool NPCC—Northeast Power Coordinating Council SERC—Southeastern Electric Reliability Council SPP—Southwest Power Pool WSCC—Western Systems Coordinating Council FPCC—Florida Power Coordinating Council.

North American Electric Reliability Organization (NAERO)—Concept that NERC could be given legal control as a non-regulated entity to formulate and enforce transmission rules, chartered by FERC.

North American Energy Standard Board (NAESB)—Private organization for the development and promotion of standards (including standard contracts and practices) for wholesale and retail natural gas and electricity transactions.

North American Industry Classification System (NAICS)—A classification scheme, developed by the Office of Management and Budget that replaced the Standard Industrial Classification (SIC) System in 1997, that categorizes establishments according to the types of production processes they primarily use.

North East Regional Transmission Organization (NERTO)—conceptual name for grouping of two or more ISOs in the Northeast (e.g., NY ISO, NE ISO, PJM, Ontario IMO); attempts to create such a grouping effectively ended in late 2002 due to widespread opposition by states and transmission owners.

Northwest Regional Transmission Association (NRTA)—A subregional transmission group within the Western Regional Transmission Association.

Notice—The Commission issues public notices for a variety of purposes, such as Commission meeting, applications for energy projects, rate cases, technical conferences, and proposed rule makings.

Notice of Proposed Rulemaking (NOPR)—A designation used by the FERC for dockets involving new regulations.

Notional—Risk management term for describing the dollar value of a single unit, such as the incremental cost of fuel to overcome an additional degree-day of heating.

NO_x—Nitrous oxides.

NPV—See Net Present Value

NRTA—See Northwest Regional Transmission Association

NUC—See Network Use Charge

Nuclear electric power (nuclear power)—Electricity generated by the use of the thermal energy released from the fission of nuclear fuel in a reactor.

Nuclear Fuel—Fissionable materials that have been enriched to such a composition that, when placed in a nuclear reactor, will support a self-sustaining fission chain reaction, producing heat in a controlled manner for process use.

Nuclear Power Plant—A facility in which heat produced in a reactor by the fissioning of nuclear fuel is used to drive a steam turbine.

NUG—See Non-utility Generator

NYCA—See New York Control Area

NYMEX—See New York Mercantile Exchange (owned by the Chicago Mercantile Exchange, a/k/a CME).

OASIS (Advancing Open Standards for the Information Society)—To aid demand response and end user demand management, a technical committee of this international organization (focused on web security, communication, etc.) was created in 2009 to "define the means of interaction between Smart Grids and their end nodes, including Smart Buildings and Facilities, Enterprises, Industry, Homes, and Vehicles. Dynamic pricing, reliability, and emergency signals must be communicated through interoperability mechanisms that meet business and energy needs, scale, use a variety of communication technologies, maintain security and privacy, and are reliable."

OASIS (Open-Access Same-Time Information System)—An electronic posting system for transmission access data that allows all Transmission Customers to view the data simultaneously, as required and defined in FERC Order 889.

OATT—See Open Access Transmission Tariff

Obligation to Serve—The obligation of a utility to provide electric service to any customer who seeks that service and is willing to pay the rates set for that service. Traditionally, utilities have assumed the obligation to serve in return for an exclusive monopoly franchise.

Off Peak—Those hours or other periods defined by contract or other agreements or guides as periods of lower electrical demand.

Offer—A motion to sell a financial product at a specified price. Also known as ask.

Off-peak Gas—Gas that is to be delivered and taken on demand when demand is not at its peak.

Offset—The elimination of a current long or short position by the opposite transaction; a sale offsets a long position; a purchase offsets a short position.

OFGEM—Office of Gas and Electricity Markets (British version of FERC).

Ohm—The unit of measurement of electrical resistance. The resistance of a circuit in which a potential difference of 1 volt produces a current of 1 ampere.

Oligopoly—A few sellers who exert market control over prices.

Once-through cooling (OTC)—process wherein water (typically from a nearby waterway) is used to cool power plants; the water is taken in, picks up generator waste heat and then immediately discharged without further use

On Peak—Those hours or other periods defined by contract or other agreements or guides as periods of higher electrical demand.

Open—The price at which the commodity or security starts a trading day.

Open access—A regulatory mandate to allow others to use a utility's transmission and distribution facilities to move bulk power from one point to another on a non-discriminatory basis for a cost-based fee.

Open Access Order No. 888—Requires utilities to allow others to use their transmission and distribution facilities, to move bulk power from one point to another on a non-discriminatory basis for a cost-based fee.

Open Access Transmission Tariff (OATT)—FERC-mandated pricing regime that ensures non-discriminatory use of transmission systems by all LSEs.

Open Architecture—Standardization, documentation, and publication of meter system parameters that allow data to be exchanged among authorized parties from an access point to the point at which data are of billing quality. Generally used in discussions of meter standards that allow any supplier of power to accept data from any meter that follows an "open architecture" specification, thus avoiding proprietary meter standards that could obsolete a customer's metering should it wish to switch to a provider using a different meter spec.

Open Interest—Total number of contracts in a commodity or options market that are still open, meaning they have not been exercised, closed out, or allowed to expire. Also called a "commitment."

Open Order—A resting order that is good until canceled.

Opening Price—The price for a given commodity generated by trading through open outcry at the opening of trading on a commodity exchange.

Operable Nuclear Unit—A nuclear unit is "operable" after it completes low power testing and is granted authorization to operate at full power. This occurs when it receives its full power amendment to its operating license from the Nuclear Regulatory Commission.

Operating Criteria—The fundamental principles of reliable interconnected systems operation.

Operating Guides—Operating practices that a Control Area or systems functioning as part of a Control Area may wish to consider. The application of Guides is optional and may vary among Control Areas to accommodate local conditions and individual system requirements.

Operating Instructions—Training documents, appendices, and other documents that explain the Criteria, Requirements, Standards, and Guides.

Operating Policies—The doctrine developed for interconnected systems operation. This doctrine consists of Criteria, Standards, Requirements, Guides, and instructions and apply to all Control Areas.

Operating Procedures—A set of policies, practices, or system adjustments that may be automatically or manually implemented by the system operator within a specified time frame to maintain the operational integrity of the interconnected electric systems.

    Automatic Operating Systems—Special protection systems, remedial action schemes, or other operating systems installed on the electric systems that require no intervention on the part of system operators.

    Normal (Pre-contingency) Operating Procedures—Operating procedures that are normally invoked by the system operator to alleviate potential facility overloads or other potential system problems in anticipation of a contingency.

    Post-contingency Operating Procedures—Operating procedures that may be invoked by the system operator to mitigate or alleviate system problems after a contingency has occurred.

Operating Requirements—Obligations of a Control Area and systems functioning as part of a Control Area. Operating Reserves:

    Spinning Reserve Service—Provides additional capacity from electricity generators that are on-line, loaded to less than their maximum output, and available to serve customer demand immediately should a contingency occur.

    Supplemental Reserve Service—Provides additional capacity from electricity generators that can be used to respond to a contingency within a short period, usually ten minutes.

Operating Reserve Demand Curve (ORDC)—ERCOT term (2013) for process used to set pricing for reserve capacity. This is not the same as ICAP, FCM, or RPM, which charge customers based on their coincident peak with an ISO. Instead, it is the pricing mechanism for seeking extra grid-wide reserve capacity.

Operating Standards—The obligations of a Control Area and systems functioning as part of a Control Area that are measurable. A Standard may specify monitoring and surveys for compliance.

Operating Transmission Limit—The maximum value of the most critical system operating parameter(s) which meets: (a) pre-contingency criteria as determined by equipment loading capability and acceptable voltage conditions, (b) transient performance criteria or, (c) post-contingency loading and voltage criteria.

Opt Out—A right of an individual end-use customer to decide not to buy from a given aggregator. Typically used in situations where one or more aggregators are identified as the primary suppliers in an area, as in the case of a standard offer, a competition for a competitive franchise, a community access entity, or a co-op.

Options—An option is a contractual agreement that gives the holder the right to buy (call option) or sell (put option) a fixed quantity of a security or commodity (for example, a commodity or commodity futures contract), at a fixed price, within a specified period of time. May either be standardized, exchange-traded, and government regulated, or over-the-counter customized and non-regulated. See also American Option, European Option.

OTC—See Over-the-Counter or Once-through Cooling

Outage—The period during which a generating unit, transmission line, or other facility is out of service. A forced outage is the removal from service availability of a generating unit, transmission line, or other facility for emergency reasons or a condition in which the equipment is unavailable due to unanticipated failure.
> **Forced Outage Rate**—The hours a generating unit, transmission line, or other facility is removed from service, divided by the

total number of hours the facility was connected to the electricity system, expressed as a percent.

Maintenance Outage—The removal of equipment from service availability to perform work on specific components that can be deferred beyond the end of the next weekend, but requires the equipment be removed from service before the next planned outage. Typically, a Maintenance Outage may occur anytime during the year, have a flexible start date, and may or may not have a predetermined duration.

Planned Outage—Removing the equipment from service availability for inspection and/or general overhaul of one or more major equipment groups. This outage usually is scheduled well in advance.

Other generation—Electricity originating from these sources biomass, fuel cells, geothermal heat, solar power, waste, wind, and wood.

Out-of-the-Money—An option that has no intrinsic value. For calls, an option whose exercise price is above the market price of the underlying future. For puts, an option whose exercise price is below the futures price.

Overbought—A technical opinion that the market price has risen too steeply and too fast in relation to underlying fundamental factors.

Overlap Regulation Service—A method of providing regulation service in which the control area providing the regulation service incorporates some or all of another control area's tie lines and schedules into its own automatic generation control/area control error equation.

Oversold—A technical opinion that the market price has declined too steeply and too fast in relation to underlying fundamental factors.

Over-the-Counter (OTC)—A virtual marketplace in which participants buy and sell bilaterally among themselves commodities based on offers posted on electronic bulletin boards, or offered by phone; differs from a public exchange wherein all pay a price posted on the exchange's board at a given time.

Pancake—The addition of multiple transmission charges ("pancaking") to the cost of power as it passes through many control areas.

Par or Basis Grade—The grade or grades specified in a given futures contract for delivery. A contract may permit deviations from the par grade subject to appropriate premiums or discounts.

Parallel Path Flow—As defined by NERC, this refers to the flow of electric power on an electric system's transmission facilities resulting from scheduled electric power transfers between two other electric systems. (Electric power flows on all interconnected parallel paths in amounts inversely proportional to each path's resistance.) During the Blackout of 2003 (in New York, Ohio, Michigan, and Canada), such flow around what was referred to as the "Lake Erie Loop" resulted in a voltage overload which caused many systems to disconnect to protect themselves.

Parol Evidence—Oral or verbal testimony, especially relating to the terms of a written agreement. In common law, it may be barred from contract cases where the evidence is extrinsic to a contract's terms and conditions.

Participating Transmission Owner (PTO)—A utility holding the controlling transmission assets in a given area, possibly as part of a transco or ISO.

Parties—Any two entities entering into a contract for services between them.

Peak Demand or Peak Load—The maximum level of electric demand in a specified time period. For billing purposes, many utilities measure the highest level of consumption during a 15 or 30-minute time period in a month and divide it by a quarter or half-hour to derive an averaged peak demand for that month. This is not to be confused with the instantaneously high (and typically brief) demand in kilowatts that occurs when a device (e.g., motor) is first started.

PBR—See Performance-Based Regulation

Peak Energy Rent (PER)—a forced reduction in capacity payments that would automatically accompany a sudden price rise in the real-time energy market. This process would deter price gouging.

Peak Load Contribution (PLC)—A benchmark measurement ensuring that a curtailment service provider is paid for following the dispatch instruction but not for anything outside of that directive. This term may also be used to describe a facility's contribution to an ISO's annual peak, know variously as a customer's "Installed Capacity (ICAP)," or "Firm Capacity Requirement (FCR)."

Peak Load Plant—A plant usually housing old, low-efficiency steam units, gas turbines, diesels, or pumped storage hydroelectric equipment normally used during the peak-load periods.

Peaking Capacity—Capacity of generating equipment normally reserved for operation during the hours of highest daily, weekly, or seasonal loads. Some generating equipment may be operated at certain times as peaking capacity and at other times to serve loads on an around-the-clock basis.

PER—See Peak Energy Rent

Percent difference—The relative change in a quantity over a specified time period. It is calculated as follows: The current value has the previous value subtracted from it; this new number is divided by the absolute value of the previous value; that new number is then multiplied by 100.

Performance-based Regulation (PBR)—Any rate-setting mechanism which attempts to link rewards (generally profits) to desired results or targets. PBR sets rates, or components of rates, for a period of time based on external indices rather than a utility's cost-of-service. PBR is a form of rate regulation which may provide utilities with better incentives to reduce their costs than occurs with cost-of-service regulation.

Petroleum coke: See Coke (petroleum).

Phasor Diagrams—Diagrams that show all vectors for three phase current and voltage.

Physical Asset—A generating plant or transmission line, or other object, as versus a financial asset that exists only in monetary form.

Pin Risk—The risk to a trader who has sold an option that, at expiration, has a strike price identical to, or pinned to, the underlying futures price.

Pipeline—A pipe through which oil, its products, or natural gas is pumped between two points, either offshore or onshore.

PJM—Name of wholesale power pool that serves most of Pennsylvania, Maryland, Delaware, and New Jersey plus Ohio, West Virginia, Virginia, portions of Kentucky, North Carolina, Indiana, and Northern Illinois; the PJM Interconnection is the official name of the ISO covering that region.

Planned Generator—A proposal by a company to install electric generating equipment at an existing or planned facility or site. The proposal is based on the owner having obtained (1) all environmental and regulatory approvals, (2) a signed contract for the electric energy, or (3) financial closure for the facility.

Planning authority (electric)—The responsible entity that coordinates and integrates transmission facility and service plans, resource plans, and protection systems.

Planning (System)—The process by which the performance of the electric system is evaluated and future changes and additions to the bulk electric systems are determined.

Planning Guides—Good planning practices and considerations that regions, subregions, power pools, or individual systems should follow. The application of planning guides may vary to match local conditions and individual system requirements.

Planning Policies—The framework for the reliability of interconnected bulk electric supply in terms of responsibilities for the development of and conformance to NERC Planning Principles and Guides and regional planning criteria or guides, and NERC and regional issue resolution processes. NERC Planning Procedures, Principles, and Guides emanate from the planning policies.

Planning Principles—The fundamental characteristics of reliable interconnected bulk electric systems and the tenets for planning them.

Planning Procedures—An explanation of how the Planning Policies are addressed and implemented by the NERC Engineering Committee, its subgroups, and the regional councils to achieve bulk electric system reliability.

Plant—A facility at which are located prime movers, electric generators, and auxiliary equipment for converting mechanical, chemical, and/or nuclear energy into electric energy. A plant may contain more than one type of prime mover. Electric utility plants exclude facilities that satisfy the definition of a qualifying facility under the Public Utility Regulatory Policies Act of 1978.

Plant Use—The electric energy used in the operation of a plant. Included in this definition is the energy required for pumping at pumped storage plants.

Plant-use Electricity—The electric energy used in the operation of a plant. This energy total is subtracted from the gross energy production of the plant; for reporting purposes the plant energy production is then reported as a net figure. The energy required for pumping at pumped storage plants is, by definition, subtracted, and the energy production for these plants is then reported as a net figure.

PLC—See Peak Load Contribution

PLC—Power Line Carrier; a mode for communicating information along power lines.

PMA—See Power Marketing Administration

POD—See Point(s) of Delivery

Point—The smallest unit of measurement of a futures price.

Point(s) of Delivery (POD)—Point(s) of interconnection on the transmission or distribution system where capacity and/or energy will be made available to the receiving party. The point(s) of delivery shall be specified in the customer's service agreement.

Point(s) of Receipt—Point(s) of interconnection on the transmission provider's transmission system where capacity and/or energy will be made available to the transmission provider by the delivering Party. This point could include an interconnection with another system or generator bus bar. The point(s) of delivery shall be specified in the service agreement.

Points on a Transmission System—Locations where power is delivered (POD), injected (POI), received (POR), or withdrawn (POW).

Point-to-point Transmission Service—The reservation and/or transmission of energy on either a firm basis and/or a non-firm basis from point(s) of receipt to point(s) of delivery, including any ancillary services that are provided by the transmission provider in conjunction with such service.

Point-to-point Transmission Service Tariff—The transmission provider's point-to-point transmission service tariff as such tariff may be amended and/or superseded from time to time.

POLR—See Provider of Last Resort—Typically, the local distribution company that provides electricity when no third-party supplier has instead been chosen.

Pool—See Power Pool

PoolCo—A specialized, centrally dispatched spot market power pool that functions as a short-term market. Some ISOs, RTOs, and power pools fit this description. It establishes the short-term market clearing price and provides a system of long-term transmission compensation contracts. It is regulated to provide open access, comparable service and cost recovery. A PoolCo provides ancillary generation services, including load following, spinning reserve, backup power, and reactive power.

POR—See Purchase of Receivables

Portfolio Manager—An EPA-sponsored energy benchmarking tool used widely to compare (on a scale of 1 to 100) the relative energy use intensity of comparible buildings in defined geographic areas.

Portfolio Requirements—Requirements on suppliers of electricity that the set of generators from which they obtain power meets certain standards. Typically refers to requirements that a minimum percentage or amount of supply be from renewable sources. Occasionally loosely applied to the more general concept of requirements or standards applying to supplier behavior.

Post Transition Ratemaking (PTR)—Regulatory efforts after the transition to full competition occurs (generally considered to start once all stranded costs have been paid off).

Postage Stamp Rates—Transmission rates based on a single charge (usually $/MWh) to move power anywhere within a given region for the same price, much as a letter may be sent for a single postage stamp charge regardless where it originated or was sent within the U.S.

Posted Price—The price a purchaser will pay for a specified product at a specified location.

POU—See Publicly Owned Utilities

Power—The rate at which energy is transferred. Electrical energy is usually measured in watts. Also used for a measurement of capacity.
    Apparent Power—The product of the volts and amperes. It comprises both real and reactive power, usually expressed in kilovolt-ampere (kVA) or megavolt-amperes (MVA).
    Reactive Power—The portion of electricity that establishes and sustains the electric and magnetic fields of alternating-current equipment. Reactive power must be supplied to most types of

magnetic equipment, such as motors and transformers. It also must supply the reactive losses on transmission facilities.

Reactive power is provided by generators, synchronous condensers, or electrostatic equipment such as capacitors and directly influences electric system voltage. It is usually expressed in kilovars (KVAR) or megavars (MVAR).

Real Power—The rate of producing, transferring, or using electrical energy, usually expressed in kilowatts (kW) or megawatts (MW).

Power Authorities—Quasi-governmental agencies that perform all or some of the functions of a public utility.

Power Exchange—A spot price pool that is governed and operated separately from the independent system operator (ISO). In a Power Exchange/ISO model, the spot price pool schedules generation and provides price bids to the ISO. The ISO may then use the sets of price bids provided by the Power Exchange to establish congestion prices, match actual demand to available supply, and facilitate the efficient short-term operation of the integrated generation and transmission system. See also Spot Price Pool.

Power Flow Program—A computerized algorithm that simulates the behavior of the electric system under a given set of conditions.

Power Marketer—Business entities engaged in buying and selling electricity. Power marketers do not usually own generating or transmission facilities. Power marketers, as opposed to brokers, take ownership of the electricity and are involved in interstate trade. These entities must file with FERC to obtain status as a wholesale power marketer.

Power Marketing Administration (PMA)—Any of five federal agencies charged with marketing power from hydroelectric plants built under federal flood control programs.

Power Pool—An association of two or more interconnected electric systems having an agreement to coordinate operations and planning for improved reliability and efficiencies.

Power production plant—All the land and land rights, structures and improvements, boiler or reactor vessel equipment, engines and engine-driven generator, turbo generator units, accessory electric equipment, and miscellaneous power plant equipment are grouped together for each individual facility.

Power Quality—The types and levels of distortions of the pure 60-cycle power waveform, such as harmonics, transients, spikes, sags, and voltage reductions. Such quality may be defined by the acceptable level of such variations as delineated in a standardized curve such as that provided by the Information Technology Industry Council (ITIC), formerly known as CBEMA.

Powder River Basin (PRB)—A region supplying low sulfur coal spanning Montana and Wyoming.

Powerhouse—A structure at a power plant site that contains the turbine and generator.

PPA—See Purchased Power Agreement

PRB—See Powder River Basin

PRD—See Price Responsive Demand

Premium—The price paid by the buyer of a commodity option to the seller of the option.

Prepayment Meters—Electric meters that allow the customer to pay a specified amount of money in advance of service to guarantee some level of minimum service while allowing low-income customers to keep within their budget. Such meters may be accompanied by a discount reflecting the lower level of service and reduced collection costs to the utility.

Price—The amount of money or consideration-in-kind for which a service is bought, sold, or offered for sale. See also Shadow Price.

Price Maker—Generator setting the market clearing price in an ISO hourly auction; often single cycle natural gas-fired plants.

Price Responsive Demand (PRD)—PJM term for end user load that may be reduced in response to dynamic pricing (e.g., critical peak pricing) or incentives.

Price Responsive Load or Price Sensitive Load—Any electrical load that may be curtailed with a user's permission, in exchange for a financial incentive from either an LDC, ISO, energy vendor, or state agency. Also called curtailable load.

Price Signals—Euphemism for changes to pricing designed to attract attention and/or action by suppliers and customers, such as demand response or time-of-use electric rates.

Price Taker—typically a baseload generator (e.g. nuclear) that bids low in an ISO hourly auction and then gets/takes the market clearing price set by others.

Prime Mover—The engine, turbine, water wheel, or similar machine that drives an electric generator; or, for reporting purposes, a device that converts energy to electricity directly (e.g., photovoltaic solar and fuel cells).

Profit—The income remaining after all business expenses are paid.

Prompt Month—The next full month about to appear.

Propane (C3H8)—A normally gaseous straight-chain hydrocarbon. It is a colorless paraffinic gas that boils at a temperature of -43.67 degrees Fahrenheit. It is extracted from natural gas or refinery gas streams. It includes all products designated in ASTM Specification D1835 and Gas Processors Association Specifications for commercial propane and HD-5 propane.

Provider of Last Resort (POLR)—Provider of electric service that is required to serve any customer requesting service in accordance with consumer protection rules and statutes. The POLR provides service to customers that do not choose an ESCO, customers that choose to leave service from an ESCO, and customers to whom an ESCO will not or can not provide service.

Pseudo-Tie—A telemetered reading or value that is updated in real time and used as a tie line flow in the Automatic Generation Control/Area Control Error equation but for which no physical tie or energy metering actually exists. The integrated value is used as a metered megawatt hour (MWH) value for interchange accounting purposes.

PTO—See Participating Transmission Owner

PTR—See Post Transition Ratemaking

Public Authority Service to Public Authorities—Public authority service includes electricity supplied and services rendered to municipalities or divisions or agencies of State or Federal governments, under special contracts or agreements or service classifications applicable only to public authorities.

Public Light and Power (PLP)—Term applied to an electricity account or usage in common areas of a facility, e.g., lobby, corridors, stairwells, rest rooms. It may be, for example, the account or bills paid by a landlord where tenants pay (either through direct or sub-metering) for their individual electricity usages in rented spaces.

Public Utility Holding Company Act of 1935 (PUHCA)—This act prohibits acquisition of any wholesale or retail electric business through a holding company unless that business forms part of an integrated public utility system when combined with the utility's other electric business. The legislation also restricts ownership of an electric business by non-utility corporations.

Public street and highway lighting—Electricity supplied and services rendered for the purpose of lighting streets, highways, parks, and other public places; or for traffic or other signal system service, for municipalities or other divisions or agencies of State or Federal governments.

Public Utility—Utility operated by a non-profit governmental or quasi-governmental entity. Public utilities include municipal utilities, cooperatives, and power marketing authorities.

Public Utility Commission (PUC)—Generic term for a state agency holding regulatory power over energy pricing, and issues related thereto.

Public Utility Regulatory Policies Act of 1978 (PURPA)—This federal statute requires States to implement utility conservation programs and create special markets for cogenerators and small power producers who meet certain standards, including the requirement that States set the prices and quantities of power the utilities must buy from such facilities. It was amended in 2005 to relax some regulations when applied in "organized" markets.

Publicly Owned Utilities (POU)—Municipal utilities (utilities owned by branches of local government) and/or co-ops (utilities owned cooperatively by customers).

PUC—See Public Utility Commission

PUHCA—See Public Utility Holding Company Act of 1935

Pumped Storage Hydroelectric Plant—A plant that usually generates electric energy during peak-load periods by using water previously pumped into an elevated storage reservoir during off-peak periods when excess generating capacity is available to do so. When additional generating capacity is needed, the water can be released from the reservoir through a conduit to turbine generators located in a power plant at a lower level.

Purchase of Receivables (POR)—Process through which a utility buys a retail supplier's contract with a customer and bills the customer for supply through the utility's delivery invoice.

Purchased Power Adjustment—A clause in a rate schedule that provides for adjustments to the bill when energy from another electric system is acquired and it varies from a specified unit base amount.

Purchased Power Agreement (PPA)—Typical name for bilateral wholesale or retail power contract.

PURPA—See Public Utility Regulatory Policy Act of 1978.

Put Option—An option which gives the buyer, or holder, the right, but not the obligation, to sell a futures contract at a specific price within a specific period of time in exchange for a one-time premium payment. It obligates the seller, or writer, of the option to buy the underlying futures contract at the designated price, should an option be exercised at that price. See also Call Option.

QF—See Qualifying Facility

QSR—Quick serve restaurant (euphemism for fast food outlet)

Qualifying Facility (QF)—A cogeneration or small power production facility that meets certain ownership, operating, and efficiency criteria established by the Federal Energy Regulatory Commission (FERC) pursuant to the Public Utility Regulatory Policies Act (PURPA). (Code of Federal Regulations, Title 18, Part 292.) Systems obtaining power through renewable sources such as wind may also qualify as QFs. Under PUHCA, such an entity is not considered a "utility," so its owner is not a "holding company."

Quark spread—The difference between the spot market values of nuclear fuel (i.e., assemblies containing uranium) and wholesale electricity at a given time, based on the conversion efficiency (Btus to produce a kWh) of a nuclear-fired plant. In essence, it is the potential savings (or cost) incurred by buying nuclear-generated power from the wholesale market instead of buying nuclear fuel to be consumed in one's own nuclear plant.

Railroad and railway electric service—Electricity supplied to railroads and interurban and street railways, for general railroad use, including the propulsion of cars or locomotives, where such electricity is supplied under separate and distinct rate schedules.

Rainbow Option—An option whose value is determined by more than one variable.

Rally—An advancing price movement following a decline.

Ramp Period—The time between ramp start and end times usually expressed in minutes, as a power plant begins daily operation, or from one kW level to another.

Ramp Rate (Schedule)—The rate, expressed in megawatts per minute, at which the interchange schedule is attained during the ramp period.

Ramsey pricing—Theoretical method for determining an ideal (and second-to-ideal) price for a product, e.g., electricity, outlined by economist Ramsey in 1927. Where a market is fully competitive and cost efficient market, price equals marginal cost. But where a natural monopoly exists, the next best choice is where price is higher than marginal cost. For electric utilities with time-of-use rates, multiple rate classes, etc., the Ramsey price yields zero profits and loss, and the difference between price and marginal cost is expressed as a percentage. Fixed costs are allocated to minimize losses to social welfare resulting from that difference.

Random Walk—See Brownian Motion

Ratable Hedge—A utility pricing method under which a portion of each forward year's MWh pricing is hedged (e.g., first year 90%, second year 60%m third year 30%) and the portions are announced publicly as a way to announce the utility's expectation of future pricing. If expected to rise, for example, the hedging percent in forward years would be lower so the utility may capture more of the upside when it occurs.

Ratchet Charge—A charge on some commercial/industrial electric bills either in addition to standard monthly demand charges or in lieu of them. It is typically a percentage (e.g. 50%) of the highest peak seen at any point during a running 12-month (or longer) period, and is charged in all months when demand is lower than that peak. It is designed to recover utility costs incurred in building and maintaining capacity used for only brief periods during a year. If peak demand drops to a lower point during the 12 months following the setting of a ratchet demand level, the ratchet charge is reset to reflect the lowest peak seen during those 12 months.

Rate—The authorized charges per unit or level of consumption for a specified time period for any of the classes of utility services provided to a customer.

Rate Base—The value of property upon which a utility is permitted to earn a specified rate of return as established by a regulatory authority. The rate base generally represents the book value of property used by the utility in providing service and may be calculated by any one or a combination of the following accounting methods: fair value, prudent investment, reproduction cost, or original cost. Depending on which method is used, the rate base includes cash, working capital, materials and supplies, and deductions for accumulated provisions for depreciation, contributions in aid of construction, customer advances for construction, accumulated deferred income taxes, and accumulated deferred investment tax credits.

Rate Case—A proceeding that involves the rates to be charged for a service that is provided by a utility.

Rate of Return (ROR)—The ratio (percentage) of profits (or earnings) compared to capital or asset value.

Rate Schedule—The rates, charges, and provisions under which service is supplied to the designated class of customers

Ratemaking Authority—A utility commission's legal authority to fix, modify, approve, or disapprove rates, as determined by the powers given the commission by a State or Federal legislature.

Rating—The operational limits of an electric system, facility, or element under a set of specified conditions.
  Continuous Rating—The rating as defined by the equipment owner that specifies the level of electrical loading, usually expressed in megawatts (MW) or other appropriate units that a system, facility, or element can support or withstand indefinitely without loss of equipment life.
  Normal Rating—The rating as defined by the equipment owner that specifies the level of electrical loading, usually expressed

in megawatts (MW) or other appropriate units that a system, facility, or element can support or withstand through the daily demand cycles without loss of equipment life.

Emergency Rating—The rating as defined by the equipment owner that specifies the level of electrical loading, usually expressed in megawatts (MW) or other appropriate units, that a system, facility, or element can support or withstand for a finite period. The rating assumes acceptable loss of equipment life or other physical or safety limitations for the equipment involved.

Ratio Utility Billing System (RUBS)—method for fairly allocating where metering is not practical or cost-effective. It may use any fair and consistent method (e.g., square footage, frontage) including proration of sub-meter readings or some other mathematical allocation method.

RCA—See Receiving Control Area

R&D—See Research and Development

Reactance—Technical term to describe voltage drop in an alternating current transmission line; analogous to resistance in a direct current transfer of power.

Reactive power—The portion of electricity that establishes and sustains the electric and magnetic fields of alternating-current equipment. Reactive power must be supplied to most types of magnetic equipment, such as motors and transformers. Reactive power is provided by generators, synchronous condensers, or electrostatic equipment such as capacitors and directly influences electric system voltage. It is a derived value equal to the vector difference between the apparent power and the real power. It is usually expressed as kilovolt-amperes reactive (KVAR) or megavolt-ampere reactive (MVAR). See Apparent Power, Power, Real Power.

Reactive Supply and Voltage Control from Generating Sources Service—Provides reactive supply through changes to generator reactive output to maintain transmission line voltage and facilitate electricity transfers.

Real Power—The component of electric power that performs work, typically measured in kilowatts (kW) or megawatts (MW)--sometimes referred to as Active Power. The terms "real" or "active" are often used to modify the base term "power" to differentiate it from Reactive Power and Apparent Power. See Apparent Power, Power, Reactive Power.

Real-time Pricing (RTP)—The pricing of electricity based on the cost of the electricity available for use at the time the electricity is demanded by the customer. As distinguished from market-based pricing, RTP may be applied to that power demand above a defined base usage for a given customer, and not to all power consumed by that customer. RTP may also encompass charges for transmission and distribution whereas market-based rates cover only the energy (and possibly capacity) portion of an electric bill.

REC—See Renewable Energy Certificate

REC Bucket—In California, one of three levels of interconnection capability, with 1 being highest, wherein a utility required to buy RECs will pay more for renewable power that is more readily accessible for its use. Bucket 3 (which may cost 1/20 of a Bucket 1 REC) is roughly comparable to RECs that are typically bought elsewhere.

REC/REMC (Rural Electric Cooperative, Rural Electric Member Cooperative)—See Cooperative Electric Utility

Recallability—The right of a transmission provider to interrupt all or part of a transmission service for any reason, including economic, that is consistent with Federal Energy Regulatory Commission policy and the transmission provider's transmission service tariffs or contract provisions.

Receipts—Purchases of fuel.

Receiving Control Area (RCA)—typically the area bounded by a utility's distribution system.

Receiving Party—The entity receiving the capacity and/or energy transmitted by the Transmission Provider to the Point(s) of Delivery.

Recourse Rate—A cost-of-service based rate for natural gas pipeline service that is on file in a pipeline's tariff and is available to customers who do not negotiate a rate with the pipeline company. Also see negotiated rate.

Reduced Emission Fuel—A combination of a fossil fuel (typically coal) with a mixture of additives designed to entrain and/or absorb pollutants such as mercury when burned with the fossil fuel. For coal, additives such as S-Sorb and MerSorb are mixed with the coal prior to combustion. The combustion pollutants are then captured in dirt filters and/or as part of fly ash, both of which are then disposed of as solid waste.

Refunctionalization—Re-labeling of transmission facilities as assets with a distribution function. Often done by utilities to gather more assets under state rate-of-return regulation and away from federal (i.e., FERC) jurisdiction.

Refund—An amount of money ordered by a regulatory agency to be returned to wholesale or retail customers after it has been determined that a rate increase has been excessive or not justified.

Region—One of the NERC Regional Reliability Councils or Affiliate.

Regional Reliability Council—One of eleven Electric Reliability Councils that form the North American Electric Reliability Council (NERC).

Regional Transmission Group (RTG)—Voluntary organization (chartered by FERC) of transmission owners, transmission users, and other entities interested in coordinating transmission planning and expansion and use on a regional and interregional basis. ISOs are considered RTGs, but RTG as a term was supplanted by Regional Transmission Organization (RTO) as a generic name that includes ISOs and transcos.

Regional Transmission Organization (RTO)—Term coined by FERC Chairman Hoecker in 1998 to describe any of several types of transmission overseers, including ISOs, ISAs, RTGs, and transcos. See also Independent System Operator (ISO) and Control Area.

Regression Modeling—Process by which a graphic trendline is developed and interpolated backward to cross either the X or Y axis (e.g., to reveal energy used at a balance point temperature).

Regulation—The government function of controlling or directing economic entities through the process of rulemaking and adjudication.

Regulation (as applied to demand response)—A service provided under contract between a provider and a grid operator designed to match, on a very short-term basis (e.g., second-by-second) supply with load. Doing so very quickly maintains voltage and supply reliability. Such a service has gained value as highly intermittent sources (mainly renewables such as wind and solar PV) are added to a grid.

Regulation and Frequency Response Service—Provides for following the moment-to-moment variations in the demand or supply in a Control Area and maintaining scheduled Interconnection frequency.

Regulation up, regulation down—regulation up involves providing energy to stabilize the grid; regulation down means absorbing excess energy from the grid for the same purpose.

Reliability—The degree of performance of the elements of the bulk electric system that results in electricity being delivered to customers within accepted standards and in the amount desired. Reliability may be measured by the frequency, duration, and magnitude of adverse effects on the electric supply. Electric system reliability can be addressed by considering two basic and functional aspects of the electric system, Adequacy and Security.

   Adequacy—The ability of the electric system to supply the aggregate electrical demand and energy requirements of the customers at all times, taking into account scheduled and reasonably expected unscheduled outages of system elements.

   Security—The ability of the electric system to withstand sudden disturbances such as electric short circuits or unanticipated loss of system elements. See also RMR (Reliability Must Run).

Reliability coordinator (electric)—The entity with the highest level of authority for the reliable operation of the Bulk Electric System, has the wide area view of the Bulk Electric System, and has the operating tools, processes and procedures, including the authority to prevent or mitigate emergency operating situations in both next-day analysis and real-time operations. The Reliability Coordinator has the purview that is broad enough to enable the calculation of Interconnection Reliability Operating Limits, which may be based on the operating parameters of transmission systems beyond any Transmission Operators vision.

Reliability Councils—Regional reliability councils were organized after the 1965 northeast blackout to coordinate reliability practices and avoid or minimize future outages. They are voluntary organizations of transmission-owning utilities and in some cases power cooperatives, power marketers, and non-utility generators. Membership rules vary from region to region. They are coordinated through the North American Electric Reliability Council (NERC). There are nine major regional councils plus the Alaska Systems Coordinating Council.

Reliability Criteria—Principles used to design, plan, operate, and assess the actual or projected reliability of an electric system.

Reliability Must Run (RMR)—Designation of a power plant whose output is needed to maintain local reliability regardless of its operating cost or market price.

Reliability Pricing Model (RPM)—A process for pricing generation capacity based on overall system reliability requirements. Using multi-year forward auctions, participants may bid capacity in the form of generation, demand response, or transmission to meet reliability needs by location and/or an ISO market.

Reliability Unit Commitment (RUC)—when a generator that has not committed to run in an energy market is nevertheless dispatched by an ISO based on the generator's power price in order to maintain reliability, especially during periods of local transmission congestion.

Renewable Energy Certificate (REC)—Represents the environmental attributes (e.g., no carbon, NO_x, SO_x, or other undesirable emissions) of 1 MWhr of power.

Renewable Identification Number (RIN)—Credit that must be purchased by petroleum marketers in the US in lieu of producing renewable fuels such as biodiesel made from vegetable oils. RINs are analogous to RECs that must be bought by electric utilities in lieu of producing or purchasing power from renewable sources under a RPS.

Renewable Natural Gas (RNG)—Methane made from accelerated decomposition of forest waste and/or animal manure via chemical or anaerobic digestive processes. The basic concept is that the carbon entrained in such waste would eventually be naturally released into the atmosphere, potentially as methane (which is a much more potent GHG than CO_2). Capturing it instead for combustion greatly reduces that impact. Because the resources for its production are continuously furnished by nature, rather than the limited supply of natural gas from underground seams, the manufactured methane is considered renewable. A small amount of RNG is already mixed into gas pipelines in some parts of the U.S.

Renewable Portfolio (or Power) Standard (RPS)—State regulatory requirement that, by a defined date, a defined percentage of generation must be supplied by renewable energy sources, such as hydroelectric, solar, wind, geothermal, or biogas.

Renewable Resources—Renewable energy resources are naturally replenishable, but flow-limited. They are virtually inexhaustible in duration but limited in the amount of energy that is available per unit of time. Some (such as geothermal and biomass) may be stock-limited in that stocks are depleted by use, but on a time scale of decades, or perhaps centuries, they can probably be replenished. Renewable energy resources include: biomass, hydro, geothermal, solar and wind. In the future they could also include the use of ocean thermal, wave, and tidal action technologies. Utility renewable resource applications include bulk electricity generation, on-site electricity generation, distributed electricity generation, non-grid-connected generation, and demand-reduction (energy efficiency) technologies.

REP—See Retail Electricity Provider

Replacement electricity—Locally generated power that supplements out-of-state renewable sources to meet delivery requirements. Replacement electricity, also called "firming and shaping" transactions, may be classified as zero GHG emissions, as long as the power plants were located in the same balancing authority as the associated renewable generation.

Request for Information (RFI)—See Request for Qualifications

Request for Proposal (RFP)—Document distributed by a customer seeking offerings and bids from suppliers of services. Sometimes also known as a Request for Quotes (RFQ).

Request for Qualifications (RFQ)—Document distributed by a customer seeking delineation of credentials for suppliers of specific types of services. Sometimes also known as a Request for Information (RFI).

Request for Quote (RFQ)—See Request for Proposal. Also called Request for Price (RFP).

Request for Rehearing or Appeal—A request of rehearing or appeal is a pleading by any party to a proceeding before a regulatory agency petitioning it to reconsider an order in that proceeding. There are statutory deadlines for filing requests for rehearing.

Rerating—A change in the capability of a generator due to a change in conditions such as age, upgrades, auxiliary equipment, cooling, etc.

Reregulation—The design and implementation of regulatory practices to be applied to the remaining regulated entities after restructuring of the vertically integrated electric utility. The remaining regulated entities would be those that continue to exhibit characteristics of a natural monopoly, where imperfections in the market prevent the realization of more competitive results, and where, in light of other policy considerations, competitive results are unsatisfactory in one or more respects. Reregulation could employ the same or different regulatory practices as those used before restructuring.

Research and Development (R&D)—Research is the discovery of fundamental new knowledge. Development is the application of new knowledge to develop a potential new service or product. Basic power sector R&D is most commonly funded and conducted through the Department of Energy (DOE), its associated government laboratories, university laboratories, the Electric Power Research Institute (EPRI), and private sector companies.

Reserve Margin (Operating)—The amount of unused available capability of an electric power system at peak load for a utility system as a percentage of total capability. Such capacity may be maintained for the purpose of providing operational flexibility and for preserving system reliability.

Reserve:
   Operating Reserve—That capability above firm system demand required to provide for regulation, load forecasting error, equipment forced and scheduled outages, and local area protection.
   Spinning Reserve—Unloaded generation, which is synchronized and ready to serve additional demand. It consists of Regulating Reserve and Contingency Reserve.

Regulating Reserve—An amount of spinning reserve responsive to Automatic Generation Control, which is sufficient to provide normal regulating margin.
   Contingency Reserve—An additional amount of operating reserve sufficient to reduce Area Control Error to zero in ten minutes following loss of generating capacity, which would result from the most severe single contingency. At least 50% of this operating reserve shall be Spinning Reserve, which will automatically respond to frequency deviation.
   Nonspinning Reserve—That operating reserve not connected to the system but capable of serving demand within a specific

time, or Interruptible Demand that can be removed from the system in a specified time. Interruptible Demand may be included in the Nonspinning Reserve provided that it can be removed from service within ten minutes.

Planning Reserve—The difference between a Control Area's expected annual peak capability and its expected annual peak demand expressed as a percentage of the annual peak demand.

> **Locational Forward Reserve (LFR)**—a form of locational capacity (and the charge for it) deployed when localized demand within a utility territory results in a sufficiently high local pricing to require startup of generators rarely used; in essence, another form of RMR; exists in New England ISO region, and ISO handles the charge for LFR.

Residential—The residential sector is defined as private household establishments which consume energy primarily for space heating, water heating, air conditioning, lighting, refrigeration, cooking, and clothes drying. The classification of an individual consumer's account, where the use is both residential and commercial, is based on principal use.

Resource Efficiency—The use of smaller amounts of physical resources to produce the same product or service. Resource efficiency involves a concern for the use of all physical resources and materials used in the production and use cycle, not just the energy input.

Resource Shuffling—To reduce a power supplier's emissions rate within a state, it sells its own coal-generated power outside its state, while importing clean energy from another state.

Response Rate
- Emergency Response Rate—The rate of load change that a generating unit can achieve under emergency conditions, such as loss of a unit, expressed in megawatts per minute (MW/Min).
- Normal Response Rate—The rate of load change that a generating unit can achieve for normal loading purposes expressed in megawatts per minute (MW/Min).

Responsible Interface Party (RIP)—person or entity interacting between an ISO and a DR provider with respect to handling payments and other activities.

Resting Order—An order away from the market, waiting to be executed.

Restricted-universe Census—This is the complete enumeration of data from a specifically defined subset of entities including, for example, those that exceed a given level of sales or generator nameplate capacity.

Restructuring—The reconfiguration of the vertically integrated electric utility. Restructuring usually refers to separation of the various utility functions into individually operated and -owned entities.

Retail—Sales covering electrical energy supplied for residential, commercial, and industrial end-use purposes. Other small classes, such as agriculture and street lighting, also are included in this category.

Retail Competition—A system under which more than one electric provider can sell to retail customers, and retail customers are allowed to buy from more than one provider. See also Direct Access.

Retail Customer—Any customer receiving power for end usage on his side of the meter, and not for redistribution/resale to others.

Retail Electricity Provider—Texas term for an unregulated provider of retail power (often called an REP). Comparable to a competitive retail energy supplier (CRES) in Ohio, an alternative retail electric supplier (ARES) in Illinois, and an energy service company (ESCo) in New York.

Retail Market—A market in which electricity and other energy services are sold directly to the end-use customer.

Retail Service Company—A company that provides the ultimate consumer of electricity with end-use services such as power, energy efficiency services, metering and billing, on-site generation, and other unbundled services.

Retail Supplier—An entity, other than the LDC, that can perform energy and customer service functions in a competitive environment, including provision of energy and assistance in the efficiency of its use.

Retail Wheeling—See Direct Access

Revenue—The total amount of money received by a firm from sales of its products and/or services, gains from the sales or exchange of assets, interest and dividends earned on investments, and other increases in the owner's equity except those arising from capital adjustments.

Revenue Decoupling Mechanism (RDM)—A tariff-based charge to pay utilities for losses due to reduced consumption from energy efficiency programs. It is designed to remove any bias against programs that otherwise reduce utility sales and revenue.

Revenue requirement—The value claimed by a utility to keep it operating properly and (in the case of investor-owned utilities) profitably.

RFI—See Request for Information

RFP—See Request for Proposal

Revenue Sufficiency Guarantee (RSG) charges—Payments made to generators that were not dispatched in a day-ahead market but are then called upon to maintain grid reliability. The charges are paid if the resulting real-time energy revenue is not enough to cover actual costs of power production. This term is used in MISO (first mentioned in 2013).

Ricochet—An Enron term for taking advantage of regional market price differentials and a lack of uniform price regulations. Prior to coining of this term by Enron personnel, this technique was known as "megawatt laundering." Similar to the Fat Boy strategy, a wholesale power marketer buys electricity at a low price (possibly capped by state regulators or the ISO) and sells it to an out-of-state customer (possibly its own subsidiary). Whereas in Fat Boy the deal would simply end, with the profit being made on the differential between the two markets' prices, under Ricochet that expensive power is then brought back into the originating state and sold to the ISO (which needs it to meet immediate demand) at a higher price, potentially generating an even higher profit.

Right-of-way (electric)—A corridor of land on which electric lines may be located. The Transmission Owner may own the land, own an easement, or have certain franchise, prescription, or license rights to construct and maintain lines.

RIN—See Renewable Identification Number

Ring Wave—a power wave surge that originates during disturbances of power flow within a building, occurring more frequently than lightning surges. The standard ring wave that must be accepted without failure by a data system or solid-state driver (e.g., LED lighting power supply) involves a 0.5 microsecond rise time and a decaying oscillation at 100 kHz. Some LED power drivers failed when experiencing such a surge.

Ringfencing—The financial and corporate isolation of a regulated public utility from the risks of unregulated activities carried out by its affiliates. Regulators ringfence in order to keep down the utility's cost of capital and to prevent the utility from buying services at above-market prices from affiliates (and thereby passing on those costs to consumers).

RIP—See Responsible Interface Party

RMR—See Reliability Must Run

Roll-over—A special straddle trading procedure involving the shift of one month of a straddle into another futures month while holding the other contract month. The shift can take place in either the long or short straddle month.

Round Trip—The practice of selling power to either an affiliate or another company and then buying it back for the same price for the sole purpose of artificially increasing trading volume to support claims of corporate growth or financial strength. This term became popular during the Enron controversy. Example: Company A might sell 100 megawatts of power to Company B on an online exchange at $10 per megawatt. Company B then turns around and sells an identical volume to Company A at the same price. No power is transferred between parties, no money is exchanged and the trades don't have any economic value—but the trades show up on each party's books as sales. Also called Wash and Back-to-Back Trading. See also Bragawatts.

Round Turn—The execution for the same customer of a purchase and sale which offset each other.

RPM—See Reliability Pricing Model

RPS—See Renewable Portfolio (or Power) Standard

RTG—See Regional Transmission Group

RTO—See Regional Transmission Organization

RTP—See Real-Time Pricing

Rules of Conduct—Rules set in advance to delineate acceptable activities by participants, particularly participants with significant market power.

Running and Quick-start Capability—The net capability of generating units that carry load or have quick-start capability. In general, quick-start capability refers to generating units that can be available for load within a 30-minute period.

Rural Electric Cooperatives—Electric cooperatives located in rural areas of the country and established and operating under rules established by Congress. See also Cooperative Electric Utility.

Sales Customer—A customer who buys natural gas or electricity with a package of other services from a utility.

Sales for Resale—Energy supplied to other electric utilities, cooperatives, municipalities, and Federal and State electric agencies for resale to ultimate consumers.

SCED—See Security-Constrained Economic Dispatch System

Schedule—An agreed-upon transaction size (megawatts), start and end time, beginning and ending ramp times and rate, and type required for delivery and receipt of power and energy between the contracting parties and the Control Area(s) involved in the transaction.

Schedule Confirmation—The process of verifying the accuracy of an interchange schedule(s) between all the entities to the transaction.

Schedule Implementation—The process of entering the details of a negotiated schedule into the control system(s) of a Control Area(s) involved in a transaction of power and energy. Scheduling, System Control, and Dispatch Service Provides for a) scheduling, b) confirming and implementing an interchange schedule with other Control Areas, including intermediary Control Areas providing transmission service, and c) ensuring operational security during the interchange transaction.

Schedule Period—The length of time between the nominal starting and ending time of each schedule.

Scheduled Losses—The scheduled power transfer to a transmission provider for compensation of losses incurred on that provider's transmission system due to a transfer of power between purchasing and selling entities.

Scheduled Outage—The shutdown of a generating unit, transmission line, or other facility, for inspection or maintenance, in accordance with an advance schedule.

Scheduling Coordinator—Often abbreviated SC, any entity responsible for gathering and reporting load profiles of many customers in order to plan generation and transmission. Such services are typically performed by utilities and large marketers, and offered by them to smaller marketers. An ISO also acts as a Scheduling Coordinator for the entire load passing through its jurisdiction.

Scoping—The scoping process is used to solicit public input on potential issues and whether there is a potential for significant adverse affects to the human environment from a proposed energy project, and identify the scope of the Environmental Assessment or Environmental Impact Statement to be prepared.

SCR—See Special Case Resources; see also Selective Catalytic Reduction.

SCUC—Security-constrained unit commitment: generators that must run in order to maintain reliable power in a given zone.

Seasonal Gaming—Scheme to take advantage of a seasonal differential between market-based power pricing that remains low during a portion of the year and regulated utility pricing that is lower during the remainder of the year by switching between these two types of suppliers.

Seat—Membership owned in an exchange. Entry to an exchange is generally through seats that are limited in number. Seats are usually held by brokerage houses.

Securitization—the issuance of government-sponsored securities to pay utilities for stranded costs at an interest rate below that available to utilities, thereby possibly decreasing the overall cost to consumers. Payoff for such securities would come from transition charges attached to all electric bills stretched out over a longer period than if utilities were to do the same over a shorter period, potentially reducing the impact seen on individual electric bills.

Securitize—The aggregation of contracts for the purchase of the power output from various energy projects into one pool which then offers shares for sale in the investment market. This strategy diversifies project risks from what they would be if each project were financed individually, thereby reducing the cost of financing. Fannie Mae performs such a function in the home mortgage market.

Security—See Reliability

Security-constrained Economic Dispatch System (SCED)—a software-based system to optimally dispatch power to all locations within an ISO or region.

Selective Catalytic Reduction (SCR)—The use of a catalyst and ammonia to reduce NO_x emissions from fossil-fueled power plants and industrial boilers.

Self-generation—A generation facility dedicated to serving a particular retail customer, usually located on the customer's premises. The facility may either be owned directly by the retail customer or owned by a third party with a contractual arrangement to provide electricity to meet some, or all of, the customer's load.

Self-service Customer (SSC)—In some deregulation schemes, a customer with monthly metered demand of 1 MW or greater that procures power without using a power marketer. Also known as a Direct Access Customer in some states.

Self-service Wheeling—Primarily an accounting policy comparable to remote net metering. An entity owns generation that produces excess electricity at one site that is used at another site(s) owned by the same entity. It is given billing credit for the excess electricity (displacing retail electricity costs minus wheeling charges) on the bills for its other sites.

Self-supply tariff—Pricing for grid-connected sites that produce on-site (typically renewable) power that is fed into the facility(s) at that site, but that do not take grid power when on-site generation is insufficient to meet demand. A typical installation may be equipped with solar panels that charge batteries which then discharge when solar power is not available or is insufficient.

SEP—See Service End Points

Service Agreement—The initial agreement and any supplements thereto entered into by the generation or transmission customer and the energy services provider.

Service Delimiter Adapters—This technology is designed especially for low-income customers to allow them to control the level of electric service they receive. For electric service, the adapter is inserted between the electric meter and the electric socket. It contains a circuit breaker which is tripped when the usage limit is exceeded. An external reset button allows the customer to restore service after cutting back on usage.

Service End Points (SEP)—Destination points for electric service most often at the meter level, but potentially also nodes.

Service List—List of participants in a regulatory proceeding.

Settlement—A generic term describing terms under which an agreement has been secured; it may apply to a cash payment made upon delivery of a commodity, or the rules under which a group of parties have agreed to operate.

Settlement Price—The official closing price of the day for each futures contract, established by the exchange as a benchmark for settling margin accounts and determining invoice price for delivery on that day.

Shadow Price—The difference between the price of power on the generation side of a constraint and the price of power on the load side of a constraint (in essence, the cost of the constraint, in $/MWhr). This is the same as the difference between the locational marginal price (LMP) at the generator and the LMP at the load.

Shift Factor—Proportion of flow across a transmission network.

Short—As in to "take a short position," or to be "short on supply," indicates a supplier has only a small (or no) amount of commodity to sell; opposite of long, which indicates the supplier has a significant amount of commodity at risk and/or to sell.

Short the Basis—The purchase of futures as a hedge against a commitment to sell in the cash or spot markets. See also Basis.

Short Ton—A unit of weight equal to 2,000 pounds (as versus a metric ton, or tonne, which is 1,000 kilograms, or about 2,204 pounds). See also Long Ton.

Shoulder Month—The time period just before or after a peak period, typically the moderately climatic months of April, May, September, and October.

SIC—See Standard Industrial Classification

Simultaneous exchange—an arranged pair of purchase and sale transactions occurring simultaneously that involve the same amount of power and number of parties, with overlapping delivery periods, but at two different locations and sometimes for different prices. The simultaneous exchange is the overlapping portion (both in volume and delivery period) of these wholesale power transactions.

Single Contingency—The sudden, unexpected failure or outage of a system facility(s) or element(s) (generating unit, transmission line, transformer, etc.). Elements removed from service as part of the operation of a remedial action scheme are considered part of a single contingency.

Six-cent Law—A New York State law, created to fulfill requirements of PURPA, which was repealed in 1992. The law required that utilities enter into contracts with qualifying cogeneration, small hydro or alternate energy facilities priced at a minimum of six cents per kilowatt-hour (for commodity only, exclusive of transmission and distribution). Prior existing contracts were grandfathered against the repeal. Other states, such as California and several New England states, had similar laws, most of which were repealed prior to deregulation in those states.

Slamming—The unauthorized change of a customer's energy supplier without the customer's knowledge or consent.

Sleeve—A financial arrangement in which energy trader A (who lacks sufficient credit with energy trader B) uses energy trader C (who still has sufficient credit with trader B) to sell to trader B by first selling his energy to trader C. In effect, trader A uses trader C as his "sleeve" to trader B and pays C a premium to "rent" his credit standing with B for the deal.

SMA—See Supply Margin Assessment

Small Power Producer (SPP)—Under the Public Utility Regulatory Policies Act (PURPA), a small power production facility (or small power producer) generates electricity using waste, renewable (water, wind, and solar), or geothermal energy as a primary energy source. Fossil fuels can be used, but renewable resource must provide at least 75 percent of the total energy input. (See Code of Federal Regulations, Title 18, Part 292.)

SMD—See Standard Market Design

SMR—Small Modular Reactor—generic term for modular nuclear reactors, generally less than 600 MW (some less than 100 MW) that could be mass produced and trucked to sites for quick installation and startup.

Solar energy—The radiant energy of the sun, which can be converted into other forms of energy, such as heat or electricity.

Southwest Regional Transmission Association (SWRTA)—A subregional RTG within WRTA.

SO$_x$—Sulfur oxides.

Spark Spread—The difference between the spot market values of natural gas and electricity at a given time, based on the conversion efficiency (Btus to produce a kWh) of a given gas-fired plant. At a given plant, one might (e.g.) burn 10,000 Btu to make a kWh. If that gas cost (e.g.) $5/MMBtu, it could make power at $.05/kWh. If the price of power at that time in the same place was $.07/kWh, the spark spread would be $.02/kWh (i.e., $20/MWhr). As the conversion efficiency becomes greater, the spread between the market value of the gas and that of power derived by burning the gas becomes wider.

Special Case Resources (SCR)—Typically generators that are run out-of-merit order due to local needs, such as voltage support, but also customer-owned generation used in demand response (DR) programs.

Special Contracts—Any contract that provides a utility service under terms and conditions other than those listed in the utility's tariffs. For example, an electric utility may enter into an agreement with a large customer to provide electricity at a rate below the tariffed rate in order to prevent the customer from taking advantage of some other option that would result in the loss of the customer's load (such as cogeneration or transmission bypass). This generally allows that customer to compete more effectively in their product market.

Speculator—An individual who trades rather than hedges in commodity futures with the objective of achieving profits through the successful anticipation of price movements.

Spinning Reserve—That reserve generating capacity running at a zero load and synchronized to the electric system.

Spot Commodity—The actual physical commodity. Sometimes called a cash commodity or actuals.

Spot Markets or Spot Purchases—Any of a number of venues in which purchases and sales, as of electricity, are made by a large number of buyers and sellers, with new transactions being made continuously or at very frequent intervals. Typically, the phrase refers to a lightly or non-regulated market in which the prices, amounts, duration and firmness of the purchases and sales are publicly known, at least shortly after the transaction is completed, if not simultaneously. Spot purchases are often made by a user to fulfill a certain portion of energy requirements, to meet unanticipated energy needs, or to take advantage of low prices.

Spot Price Pool—A neutral and independent organization with no interest in generating facilities that provides an open access spot market for power. A spot price pool typically accepts hourly or half-hourly price bids no more than a day in advance. Suppliers are selected on the basis of economic dispatch taking into consideration price bids, congestion and other transmission costs. Transactions in the pool, as in any competitive market, are settled at market clearing prices or the bid of the highest priced generator scheduled to deliver power in each time period and major area in the transmission system. Spot price pools, whether voluntary or mandatory, are designed to co-exist with and facilitate markets in bilateral contracts. See also: PoolCo; Power Exchange.

Spot purchases—A single shipment of fuel or volumes of fuel purchased for delivery within 1 year. Spot purchases are often made by a user to fulfill a certain portion of energy requirements, to meet unanticipated energy needs, or to take advantage of low-fuel prices.

SPP—See Small Power Producer

SSC—See Self-Service Customer

SSR—See System Support Reliability

Stability—The ability of an electric system to maintain a state of equilibrium during normal and abnormal system conditions or disturbances.

Small-signal Stability—The ability of the electric system to withstand small changes or disturbances without the loss of synchronism among the synchronous machines in the system.

Transient Stability—The ability of an electric system to maintain synchronism between its parts when subjected to a disturbance of specified severity and to regain a state of equilibrium following that disturbance

Stability Limit—The maximum power flow possible through some particular point in the system while maintaining stability in the entire system, or the part of the system, to which the stability limit refers.

Stable Prices—Prices that do not vary greatly over short time periods. Different customers value stability in different ways. Residential and small business customers typically prefer to have prices that do not vary more frequently than annually, or at most quarterly. Very large customers may find changing hourly spot prices to be "stable" enough for their uses. See also Variable Prices.

Standard Deviation—Mathematical result equal to the square root of the mean of the squares of the deviations of a series of numbers from their arithmetic mean.

Standard Industrial Classification (SIC)—A set of codes developed in 1937 by the federal government to categorize businesses into groups with similar economic activities. In 1992, the federal Office of Management and Budget (OMB) revised SIC into the North Amerian Industrial Classification System (NAICS). It became effective in 1997.

Standard Market Design (SMD)—FERC term for the template it created in 2002, to re-arrange existing wholesale power markets into a more common system by applying "best practices" and other improvements such as LMP, CRR, etc. in order to engender various attributes FERC believed are most desirable for competitive power pricing. Those attributes include (among others) a well functioning Regional Transmission Organization (RTO), incentives for construction of new merchant power plants and transmission lines, and processes for resolving inter-regional problems, such as seams between RTOs. After about 2 years, FERC dropped this overall effort and sought to bring about some of the goals via various orders and settlements.

Standard Offer—A regulated utility power price for the commodity portion of electricity during a transition period to competition in generation supply. Usually proposed for the stated purpose of giving "customers who choose not to choose" the option of remaining with their existing supplier of electricity, and often used as a benchmark against supply-only pricing from competitive retail power suppliers.

Standby Facility—A facility that supports a utility system and is generally running under no load. It is available to replace or supplement a facility normally in service.

Standby Service—Support service that is available, as needed, to supplement a consumer, a utility system, or to another utility if a schedule or an agreement authorizes the transaction. The service is not regularly used.

Statutes—The laws passed by a government that give it the power to regulate.

Steam Electric Plant (Conventional)—A plant in which the prime mover is a steam turbine. The steam used to drive the turbine is produced in a boiler where fossil fuels are burned.

Stochastic—Approximately synonymous with "random statistical probability."

Stocks—A supply of fuel accumulated for future use. This includes coal and fuel oil stocks at the plant site, in coal cars, tanks, or barges at the plant site, or at separate storage sites.

Stop-Loss—A resting order designed to close out a losing position when the price reaches a level specified in the order. It becomes a market order when the "stop" price is reached.

Storage—Energy transferred from one entity to another entity that has the ability to conserve the energy (i.e., stored as water in a reservoir, electricity in a battery, etc.) with the intent that the energy will be returned at a time when such energy is more usable to the original supplying entity. A common synonym for Storage is Energy Banking.

Storage Facility—Underground storage of natural gas in natural geologic reservoirs such as depleted oil or gas reservoirs or natural underground caverns. Gas is transported from producing fields during periods of low demand, stored in underground storage, and then withdrawn for distribution during periods of peak demand.

Straddle (futures)—Also known as a spread, the purchase of one futures month against the sale of another futures month of the same commodity A straddle trade is based on a price relationship between the two months and a belief that the "spread" or difference in price between the two contract months will change sufficiently to make the trade profitable.

Straddle (options)—The purchase or sale of both a put and a call having the same strike price and expiration date. The buyer of a straddle benefits from increased volatility and the seller benefits from decreased volatility.

Strandable Benefit(s)—A benefit that becomes stranded when an industry is restructured without providing for the continued delivery of this public good or service.

Stranded Assets/Stranded Costs—See Embedded Costs Exceeding Market Prices.

Stranded Benefits—Public interest programs and goals which could be compromised or abandoned by a restructured electric industry. These potential stranded benefits might include environmental protection, fuel diversity, energy efficiency, low-income ratepayer assistance, and other types of socially beneficial programs.

Stranded costs—Costs incurred by a utility which may not be recoverable under market-based retail competition. Examples include undepreciated generating facilities, deferred costs, and long-term contract costs.

Strangle—An options position consisting of the purchase or sale of put and call options having the same expiration but different strike prices.

Strike Price—The price at which the underlying futures contract is bought or sold in the event an option is exercised. Also called an exercise price. Also, a price accepted by your firm prior to its availability, such that it may be communicated to an energy vendor so that a deal securing it may be consummated as soon as that price appears.

Strip—The simultaneous purchase (or sale) of futures positions in consecutive months.

Subbituminous coal—A coal whose properties range from those of lignite to those of bituminous coal and used primarily as fuel for steam-electric power generation. It may be dull, dark brown to black, soft and crumbly, at the lower end of the range, to bright, jet-black, hard, and relatively strong, at the upper end. Subbituminous coal contains 20 to 30 percent inherent moisture by weight. The heat content of subbituminous coal ranges from 17 to 24 million Btu per ton on a moist, mineral-matter-free basis. The heat content of subbituminous coal consumed in the United States averages 17 to 18 million Btu per ton, on the as-received basis (i.e., containing both inherent moisture and mineral matter).

Subregion—A portion of a Region. A subregion may consist of one or more Control Areas.

Substation—A facility for switching electrical elements, transforming voltage, regulating power, or metering.

Sulfur—A yellowish nonmetallic element existing at various levels of concentration in fossil fuels whose combustion releases sulfur compounds considered harmful to the environment. Coal is classified as being low-sulfur at concentrations of 1 percent or less or high-sulfur at concentrations greater than 1 percent.

Sunk Cost—In economics, a sunk cost is a cost that has already been incurred, and therefore cannot be avoided by any strategy going forward.

Sunshine Notice—A legal notice that is required under the Sunshine Act. This Act states that public notice must be published at least one week in advance of all meetings where a quorum of an agency officials will be conducting or deciding official agency business.

Subject Matter Expert (SME)—business-speak for the in-house person who knows a lot about a subject.

Superconducting Magnetic Energy Storage (SMES)—process through which electricity is stored in a superconducting magnet that is maintained in a cryogenic state to almost fully eliminate efficiency losses.

Supervisory Control And Data Acquisition (SCADA)—A utility term for system-wide metering and data handling system; being adopted by some large end users to describe their own metering and load control systems. A system of remote control and telemetry used to monitor and control the electric system.

Supplemental gaseous fuels supplies—Synthetic natural gas, propane-air, coke oven gas, refinery gas, biomass gas, air injected for Btu stabilization, and manufactured gas commingled and distributed with natural gas.

Supplier—See Retail Supplier

Supply Margin Assessment (SMA)—A FERC term coined in 2002 to describe a determination of sufficient existing excess generating capacity in a geographic area (e.g., that of an ISO or RTO) to maintain reliable supply. This internal analysis is seen as an interim step toward FERC's Standard Market Design (SMD) which calls for a minimum of 12%-18% of capacity above forecast peaks to ensure reliability and competitive wholesale regional power markets.

Supply-side—Activities conducted on the utility's side of the customer meter. Activities designed to supply electric power to customers, rather than meeting load though energy efficiency measures or on-site generation on the customer side of the meter.

Surge—A transient variation of current, voltage, or power flow in an electric circuit or across an electric system.

Suspended Rates—New rates that have been accepted for review by FERC but not made effective for a period of time, up to a maximum period of five months.

Sustained Orderly Development—A condition in which a growing and stable market is identified by orders that are placed on a reliable schedule. The orders increase in magnitude as previous deliveries and engineering and field experience lead to further reductions in costs. The reliability of these orders can be projected many years into the future, on the basis of long-term contracts, to minimize market risks and investor exposure. See also Commercialization.

SVC—Static VAR compensator: electric device designed to maintain system balance.

Swap—A financial trade involving the exchange of two different pricing structures between users of a commodity. For example, Consumer A (who normally purchases a fixed priced commodity) and Consumer B (a buyer of a variably priced commodity) would (through a broker) agree to pay for each other's purchases. Consumer A thus assumes the risk of a variably priced commodity while potentially seeing a lower annual bill, while Consumer B gets price assurance while risking a potentially higher annual bill. Swaps can be conducted directly by two counter-parties, or through a third party such as a bank or brokerage house.

Swing Option—A purchasing regime under which a customer's price is either constant or otherwise defined across a percent range above and below an expected usage, on an hourly, daily, monthly, and/or annual basis, before incurring a penalty or undefined price. See also Balancing and Settlement.

Switching Station—Facility equipment used to tie together two or more electric circuits through switches. The switches are selectively arranged to permit a circuit to be disconnected, or to change the electric connection between the circuits.

SWRTA—See The Southwest Regional Transmission Administration.

Synchronize—The process of connecting two previously separated alternating current apparatuses after matching frequency, voltage, phase angles, etc. (e.g., paralleling a generator to the electric system).

Synchronous Condensing—Process for maintaining voltage stability on a grid, involving use of a standby generator turbine running as a very large electric motor load. When reactive power on a grid is leading, the unit acts to retard it. When reactive power is lagging, the unit acts to excite it. Such systems may be deployed at the ends of long radial transmission lines.

Synthetic Futures—A position created by combining call and put options.

System—An interconnected combination of generation, transmission, and distribution components comprising an electric utility, an electric utility and independent power producer(s) (IPP), or group of utilities and IPP(s).

System Benefits Charge—Any of a number of non-bypassable charges imposed to collect funds to cover the above-market costs of providing public goods (system benefits) that otherwise would be stranded.

System Integration (of new technologies)—The successful integration of a new technology into the electric utility system by analyzing the technology's system effects and resolving any negative impacts that might result from its broader use.

System Operator—An individual at an electric system control center whose responsibility it is to monitor and control that electric system in real time.

System Support Reliability—MISO version of reliability-must-run, wherein some power plants are paid to run to maintain grid reliability, even if not requested to do so based on their bid pricing,

Tagging—A NERC requirement to identify interchange transactions between control areas; some have questioned the fairness of this process because it may openly reveal competitive positions or allow utilities to limit competitors' transmission options.

Take a Position—Marketing term meaning purchasing and offering of a commodity for resale.

Take-and-pay (TAP)—Form of PPA or gas purchasing agreement in which a customer commits to buy a defined volume of energy during a defined time period (i.e., term); if the seller fails to deliver a portion of the energy, the buyer pays only for what he receives.

Take-or-pay (TOP)—Form of PPA or gas purchasing agreement in which a customer commits to buy a defined volume of energy during a defined time period (i.e., term); if the buyer fails to take title to the energy, he pays as much (or a portion) of the cost of the energy as though it had been delivered.

Taking—Reducing the value of someone's property through government action without just compensation.

TAP—See Take-And-Pay

Tariff—A document, approved by the responsible regulatory agency, listing the terms and conditions, including a schedule of prices, under which utility services will be provided.

TAT—See Thermally Activated Technologies

TCC—See Transmission Congestion Contract

Technical Analysis—An approach to forecasting commodity prices which examines patterns of price changes, rates of change, and changes in trading volume and open interest, without regard to underlying fundamental market conditions.

Tender—European/Canadian term for a request for bids.

Telemetering—The process by which measurable electrical quantities from substations and generating stations are instantaneously transmitted using telecommunication techniques.

Terawatt hour—One trillion watt-hours.

Term—The duration, starting, and ending points of a contract.

Term Desk—A trading group that handles trades covering months. See also Cash Desk.

Term-Limit Pricing (TLP)—An agreement on price between a supplier and a wholesaler or jobber that runs for a specific length of time.

Theoretical Value—An option's value generated by a mathematical model given certain prior assumptions about the term of the option, the characteristics of the underlying futures contract, and the prevailing interest rates.

Therm—100,000 British thermal units.

Thermal Rating—The maximum amount of electrical current that a transmission line or electrical facility can conduct over a specified time period before it sustains permanent damage by overheating or before it violates public safety requirements.

Thermally Activated Technologies (TAT)—Central plant equipment, such as absorption chillers, that uses heat to produce cooling or other services commonly used in buildings.

Theta—The sensitivity of an option's value to a change in the amount of time to expiration.

Through-and-out Tariff—Wholesale power tariffs maintained by utilities which levy charges for moving power across or into their transmission networks from others' networks; in July 2003, FERC determined that several such tariffs in the Midwest were blocking interstate commerce and ordered them to be changed or rescinded.

Tick—The smallest monetary unit in which the movement of price of a given commodity may he expressed in futures trading. See Point.

Tie Line—A circuit connecting two or more Control Areas or systems of an electric system.

Tie Line Bias—A mode of operation under automatic generation control in which the area control error is determined by the actual net interchange minus the biased scheduled net interchange.

Tight Pool—Power pool within which a central operator dispatches generation minute-by-minute, often automatically (as versus pools that instead plan generation strategies in a more general and voluntary fashion)

Time Error—An accumulated time difference between Control Area system time and the time standard. Time error is caused by a deviation in Interconnection frequency from 60.0 Hertz.

Time Error Correction—An offset to the Interconnection's scheduled frequency to correct for the time error accumulated on electric clocks.

Time-of-day Service—See Time-of-Use (TOU) Rates and Time-Varying Pricing (TVP).

Time-of-use (TOU) Rates—The pricing of electricity based on the estimated cost of electricity during a particular time block. Time-of-use rates are usually divided into three or four time blocks per twenty-four hour period (on-peak, mid-peak, off-peak and sometimes super off-peak) and by seasons of the year (summer and winter). Real-time pricing differs from TOU rates in that it is based on actual (as opposed to tariff) prices which may fluctuate many times a day and are weather-sensitive, rather than varying with a fixed schedule.

Time-varying Pricing (TVP)—Any of the various ways to adjust power pricing based on time, including TOU, CPP, and other options for charging higher $/kWh rates when grids are stressed and/or more expensive peaking power plants must be used.

TLP—See Term-limit pricing

TLR—Transmission loading relief: a process designed to allow transmission operators to drop or change scheduled power deliveries to accommodate limitations on transmission lines; a continuing bone of contention between regulated utilities (who own most power lines) and unregulated and/or wholesale power suppliers and marketers.

Tolling—Marketing tactic in which a holder of natural gas purchases temporary spare generating capacity to convert his fuel into electricity for resale as power. Also known as Form Value Trading.

TOP—See Take-Or-Pay

Top-down—Typical approach taken when unbundling rates, and the opposite of a bottom-up approach. Top-down implies examination of initially only the energy component of a utility rate, exclusive of transmission, distribution, customer service, metering, marketing, taxes, etc. The net result is that only the energy component is subjected to competition, instead of all aspects of the utility rate. A bottom-up approach involves an examination in the reverse order and should therefore be more comprehensive. See also Load Forecast for unrelated use of Bottom-up and Top-Down.

Torrefaction—thermal process for converting biomass (e.g., wood waste) into "biocoal," a renewable form of boiler fuel. The biomass is heated (to about 500—600 F) in an oxygen-free environment to drive off moisture and volatiles, leaving an unburned char that is then compressed into pellets or briquettes. The biocoal may then be either mixed with or fully replace coal in standard coal-fired power plants, thus reducing their net emissions.

Total Transfer Capability (TTC)—The amount of electric power that can be transferred over the interconnected transmission network in a reliable manner.

Tranche (or tranch)—European term for a bid covering a portion of a customer's energy requirements; tranches may have overlapping terms so that several together satisfy an overall requirement.

Transco—See Transmission Company

Transfer Capability—The measure of the ability of interconnected electric systems to move or transfer power in a reliable manner from one area to another over all transmission lines (or paths) between those areas under specified system conditions. The units of transfer capability are in terms of electric power, generally expressed in megawatts (MW). In this context, "area" may be an individual electric system, power pool, Control Area, subregion, or NERC Region, or a portion of any of these. Transfer capability is directional in nature. That is, the transfer capability from "Area A" to "Area

B" is not generally equal to the transfer capability from "Area B" to "Area A."

Transformer—An electrical device for changing the voltage of alternating current.

Transition Costs—See Embedded Costs Exceeding Market Prices

Transmission (electric)—An interconnected group of lines and associated equipment for the movement or transfer of electric energy between points of supply and points at which it is transformed for delivery to customers or is delivered to other electric systems.

Bulk Transmission—A functional or voltage classification relating to the higher voltage portion of the transmission system.

Sub-transmission—A functional or voltage classification relating to the lower voltage portion of the transmission system.

Transmission Company—A privately owned firm whose primary asset and focus is the high-voltage transmission of power. Often called a transco.

Transmission Congestion Contract (TCC)—A financial instrument that allows a transmitter of power to define his transmission costs irrespective of the immediate cost of transmission seen on an ISO's OASIS. TCCs enable energy buyers and sellers to hedge transmission price fluctuations. A TCC holder has the right to collect or the obligation to pay congestion rents in the day-ahead market for energy associated with transmission between specified points of injection and withdrawal. Also known as Congestion Revenue Rights (CRR) and Financial Congestion Rights. When the price of a TCC is negative, it means an ISO pays the buyer the clearing price for accepting it. This occurs when congestion is expected in the direction opposite to the way a TCC is labeled. TCC auctions are typically held monthly. (FCR). See also Firm Transmission Rights (FTR).

Transmission constraint relaxation parameter—price at which ISO software may drop transmission constraints in real-time instead of continuing to redispatch resources to reduce congestion. This process is designed to minimize uplift charges at the ISO level due to apparent transmission congestion.

Transmission Constraints—Limitations on a transmission line or element that may be reached during normal or contingency system operations.

Transmission Customer—Any eligible customer (or its designated agent) that can or does execute a transmission service agreement or can or does receive transmission service.

Transmission line (electric)—A system of structures, wires, insulators and associated hardware that carry electric energy from one point to another in an electric power system. Lines are operated at relatively high voltages varying from 69 kV up to 765 kV, and are capable of transmitting large quantities of electricity over long distances.

Transmission operator (electric)—The entity responsible for the reliability of its localized transmission system, and that operates or directs the operations of the transmission facilities.

Transmission owner (electric)—The entity that owns and maintains transmission facilities.

Transmission Provider—Any public utility that owns, operates, or controls facilities used for the transmission of electric energy in interstate commerce.

Transmission Reliability Margin (TRM)—That amount of transmission transfer capability necessary to ensure that the interconnected transmission network is secure under a reasonable range of uncertainties in system conditions. See Available Transfer Capability.

Transmission Service Provider (electric)—The entity that administers the transmission tariff and provides Transmission Service to Transmission Customers under applicable transmission service agreements.

Transmission System (Electric)—An interconnected group of electric transmission lines and associated equipment for moving or transferring electric energy in bulk between points of supply and points at which it is transformed for delivery over the distribution system lines to consumers, or is delivered to other electric systems.

Transmission-dependent Utility—A utility that relies on its neighboring utilities to transmit to it the power it buys from its suppliers. A utility without its own generation sources, dependent on another utility's transmission system to get its purchased power supplies.

Transmitting Utility (Transco)—This is a regulated entity which owns, and may construct and maintain, wires used to transmit wholesale power. It may or may not handle the power dispatch and coordination functions. It is regulated to provide non-discriminatory connections, comparable service and cost recovery. According to EPAct, any electric utility, qualifying cogeneration facility, qualifying small power production facility, or Federal power marketing agency which owns or operates electric power transmission facilities which are used for the sale of electric energy at wholesale. See also Generation Dispatch & Control; Power Pool.

Transportation Customer—A customer who uses a utility's pipeline and distribution system but buys natural gas from a different supplier.

Trenwa—Precast concrete trench boxes designed to protect and cover underground high-voltage transmission lines, typically filled with insulating materials surrounding such cables.

TRM—See Transmission Reliability Margin

TTC—See Total Transfer Capability (TTC)

Turbine—A machine for generating rotary mechanical power from the energy of a stream of fluid (such as water, steam, or hot gas). Turbines convert the kinetic energy of fluids to mechanical energy through the principles of impulse and reaction, or a mixture of the two.

TVP—See Time-Varying Pricing.

TWACS—Two-Way Automatic Communications System for sending metering data and operating commands over power lines.

TX—Occasional abbreviation for transmission.

UCAP—See Unconstrained Capacity, under Capacity

UFLS—See Under-Frequency Load Shedding

ULSD—ultra low-sulfur diesel, i.e., less than .3% sulfur. In some areas, required for use with diesel generators and some diesel-fired equipment.

Ultimate customer—A customer that purchases electricity for its own use and not for resale.

Unaccounted for Energy (UFE)—See Line Loss

Unbundling—Disaggregating electric utility service into its basic components and offering each component separately for sale with separate rates for each component, e.g., generation, transmission and distribution could be unbundled and offered as discrete services.

Under-frequency Load Shedding (UFLS)—To prevent cascading blackouts, UFLS is a last-ditch measure that grid and power plant operators take to reduce the likelihood of a total or partial grid system collapse. Operating the power grid requires that nearly an equal amount of generation is put on the grid compared to what is being taken off by customers, or load.

The sudden loss of a power plant unit can undo this balance and cause system transmission frequency to deviate from where it needs to be, which can lead to more power plants tripping offline. If grid operators cannot make up for this loss of generation with so-called spinning reserves, or if the frequency drops too fast, grid operators would initiate automatic load shedding under a North American Electric Reliability Corp.-required UFLS program.

Uniform System of Accounts—Prescribed financial rules and regulations established by the Federal Energy Regulatory Commission for utilities subject to its jurisdiction under the authority granted by the Federal Power Act.

Unit—More than one class of securities traded together on the exchanges. For example, one common stock and one warrant can be traded together as a unit.

Unit Commitment—The process of determining which generators should be operated each day to meet the daily demand of the system.

Unit Seriotum—To start or operate generators, transmission lines, or other components in a pre-determined order of succession.

Universal Service—Electric service sufficient for basic needs (an evolving bundle of basic services) available to virtually all members of the population regardless of income.

Uplift—British term adopted in the US to describe payments made to an electricity supplier to cover its generating cost when market-clearing prices are insufficient to do so. Also called "make-whole" payments, they may cover costs for additional ancillary services or differences between forecasted and actual power demand. Uplift payments have also been made to maintain generators providing reserves for reliability and/or voltage support.

Uprate—An increase in available electric generating unit power capacity due to a system or equipment modification. An uprate is typically a permanent increase in the capacity of a unit.

Useful Thermal Output—The thermal energy made available for use in any industrial or commercial process, or used in any heating or cooling application, i.e., total thermal energy made available for processes and applications other than electrical generation.

UTC—Up-to-congestion; a form of wholesale power contract that does not require payment for some components of typical contracts, such as transmission reservation and balancing operating reserves. As such, they are useful as a financial, rather than physical, transaction.

Utility—A regulated entity which exhibits the characteristics of a natural monopoly. For the purposes of electric industry restructuring, "utility" refers to the regulated, vertically integrated electric company. Transmission utility refers to the regulated owner/operator of the transmission system only. Distribution utility refers to the regulated owner/operator of the distribution system which serves retail customers.

Utility distribution companies—The entities that will continue to provide regulated services for the distribution of electricity to customers and serve customers who do not choose direct access. Regardless of where a consumer chooses to purchase power, the customer's current utility, will deliver the power to the consumer.

Utility-earned Incentives—Monies paid to the utility for achievement in consumer participation in DSM programs. They are intended to influence the utility's consideration of DSM as a resource option by addressing cost recovery, lost revenue, and profitability.

VACAR—stands for Virginia And the Carolinas, a sub-region of the Southeast Reliability Council (SERC), a NERC transmission reliability region; this is also a wholesale power pricing point.

Value of lost load (VOLL)—The maximum price electricity customers would pay to avoid a power outage. Typically expressed as $/MW-hr, it was estimated to be $6,700/MW-hr after the 2011 outages and price spikes seen in ERCOT. In 2003, GDF Suez proposed raising it to $18,000 or $25,000. It is sometimes used as a benchmark for prices paid for demand response that may be able to avoid an outage.

Value-at-risk (VAR)—Analytical method for assessing how much of an asset could be impacted by a sudden change in energy pricing or availability. While useful when dealing primarily with financial instruments and transactions, energy marketers have found that this method is not as effective when handling physical transactions. A competing concept called "profit-at-risk" (PAR) has been gaining favor to handle volatility in that arena.

Value-driven Aggregator—An aggregator created to arrange supplies of electricity and related services from supplies in a manner consistent with a set of values, particularly including non-price criteria. The aggregator can be a co-op, a municipality administering a competitive franchise, or a community access entity.

Vanilla—A standard or common contractual arrangement; opposite of Exotic.

VAR—See Value-at-Risk

VAR—Volt-ampere reactive: a technical term for describing one aspect of power as it is transmitted.

Variable energy resource (VER)—Power plants (typically wind and solar) whose output is inconsistent. VERs may be charged by grid operators or utilities to which they are connected for ancillary services, e.g., regulation and frequency response, to correct for such inconsistency.

Variable Frequency Transformer (VFT)—Utility level transformer that may boost a transmission line's wattage capacity.

Variable Prices—Prices that vary frequently. Prices which are not stable. See Stable Prices.

Vega—The sensitivity of an options value to a change of volatility.

Vertical Integration—An arrangement whereby the same company owns all the different aspects of making, selling, and delivering a product or service. In the electric industry, it refers to the historically common arrangement whereby a utility would own its own generating plants, transmission system, and distribution lines to provide all aspects of electric service.

VFT—See Variable Frequency Transformer

Vintage—As applied to a renewable energy certificate (REC), its vintage is the date that the electric generation associated with the REC was measured by the grid operator or utility meter at the generator site.

Volatility—While often used generically to describe the degree of variation in pricing of a commodity, this term also has a specific mathematical definition (often abbreviated with the Greek symbol for sigma) defined as the standard deviation of a price during a defined time period (typically several months or a year).

Volatility Index—The standard deviation of hourly power prices on a monthly basis.

Volatility Smile—A type of price curve occurring when high prices drop and then rapidly rise again.

Voltage Collapse—An event that occurs when an electric system does not have adequate reactive support to maintain voltage stability. Voltage Collapse may result in outage of system elements and may include interruption in service to customers.

Voltage Control—The control of transmission voltage through adjustments in generator reactive output and transformer taps, and by switching capacitors and inductors on the transmission and distribution systems.

Voltage Limits:
• **Normal Voltage Limits**—The operating voltage range on the interconnected systems that is acceptable on a sustained basis.
• **Emergency Voltage Limits**—The operating voltage range on the interconnected systems that is acceptable for the time sufficient for system adjustments to be made following a facility outage or system disturbance.

Voltage Reduction—Any intentional reduction of system voltage by 3 percent or greater for reasons of maintaining the continuity of service of the bulk electric power supply system.

Voltage Stability—The condition of an electric system in which the sustained voltage level is controllable and within predetermined limits.

Volumetric Wires Charge—A type of charge for using the transmission and/or distribution system that is based on the volume of electricity that is transmitted.

Warrants—Type of security usually issued together with a bond or preferred stock that entitles the holder to buy a proportionate amount of common stock at a specified price, usually higher than the market price at the time of issuance, for a period of time. In essence, a form of future. Warrants are freely transferable and are traded on major exchanges.

Wash—See Round Trip

Waste coal—Usable material that is a byproduct of previous coal processing operations. Waste coal is usually composed of mixed coal, soil, and rock (mine waste). Most waste coal is burned as-is in unconventional fluidized-bed combustors. For some uses, waste coal may be partially cleaned by removing some extraneous noncombustible constituents. Examples of waste coal include fine coal, coal obtained from a refuse bank or slurry dam, anthracite culm, bituminous gob, and lignite waste.

Waste oils and tar—Petroleum-based materials that are worthless for any purpose other than fuel use.

WATSCO—The Western Association for Transmission System Coordination.

Watt—The basic unit of electrical demand or power, equal to the rate of energy transfer equal to one ampere flowing due to an electrical force, or pressure, of one volt at a power factor of 1.0.

Watt-hour (Wh)—An electrical energy unit of measure equal to 1 watt of power supplied to, or taken from, an electric circuit steadily for 1 hour.

WDG—See Wholesale Distributed Generation

Weather Derivative—A financial arrangement based on defined climatic conditions, such as dry bulb temperature or precipitation; used mainly to control volume risk; some have referred to such arrangements as "weather insurance."

Well—An opening in the ground made by drilling, boring, or any other manner, from which oil or gas (or water) is obtained. Wells may also be used to inject oil, gas, water, or other fluids into the ground.

Wellhead—Point at which natural gas is withdrawn from the ground.

Wellhead Price—What it costs to produce natural gas at the wellhead. The wellhead price does not include any charges for treating, gathering, processing, transporting, or distributing.

Wheeling—The transmission of electricity by an entity that does not own or directly use the power it is transmitting. Wholesale wheeling is used to indicate bulk transactions in the wholesale market, whereas retail wheeling allows power producers direct access to retail customers. This term is often used colloquially as meaning transmission.

Wholesale Competition—A system whereby a distributor of power would have the option to buy its power from a variety of power producers, and the power producers would be able to compete to sell their power to a variety of distribution companies.

Wholesale Customer—Any entity eligible under FERC (and some state) regulations to receive power directly from producers for redistribution to retail customers.

Wholesale Distributed Generation (WDG)—20 MW and smaller renewable energy projects that are interconnected with the distribution grid, not the transmission grid. Popular in some European countries (e.g., Germany) and parts of Canada.

Wholesale Power Market—The purchase and sale of electricity from generators to resellers (who sell to retail customers) along with the ancillary services needed to maintain reliability and power quality at the transmission level. Wholesale Sales Energy supplied to other electric utilities, cooperatives, municipals, and Federal and State electric agencies for resale to ultimate consumers.

Wholesale sales—Energy supplied to other electric utilities, cooperatives, municipals, and Federal and state electric agencies for resale to ultimate consumers.

Wholesale Storage Load (WSL)—Energy (separately metered from all other facilities) used to charge a technology that is capable of storing energy and releasing it at a later time to generate electricity. It includes losses occurring in the energy conversion process. WSL includes the following technologies: batteries, flywheels, compressed air energy storage, pumped hydroelectric power, and electrochemical capacitors.

Wholesale Transmission Services—The transmission of electric energy sold, or to be sold, at wholesale in interstate commerce (from EP-Act).

Wind energy—Kinetic energy present in wind motion that can be converted to mechanical energy for driving pumps, mills, and electric power generators

Wire House—A brokerage operating a private "wire" to its own branch offices or to other firms. See also Commission House.

Wires and Pipes—A way to refer to a utility that only distributes power and gas

Wires Charge—A broad term which refers to charges levied on power suppliers or their customers for the use of the transmission or distribution wires.

Wiresco—Another name for a transco/disco, or disco; a utility devoid of generation.

Working Gas—Gas in an underground storage facility which is available to be withdrawn and may be sold. This is in contrast to base gas, which is needed to operate the storage facility, and can't be recovered during a normal working cycle.

Writer—The seller of an option. Also known as the grantor of an option.

WRTA—The Western Regional Transmission Association, an RTG.

WSCC (The Western System Coordinating Council)—A voluntary industry association created to enhance reliability among western utilities.

WSL—See Wholesale Storage Load

WSSP (The Western Systems Power Pool)—A FERC-approved industry institution that provides a forum for short-term trades in electric energy, capacity, exchanges and transmission services. The pool consists of approximately 50 members and serves 22 states, a Canadian province and 60 million people. The WSSP is headquartered in Phoenix, Arizona.

Year-over-year (Y-o-Y)—Description of comparison of data from two adjacent years.

Yieldco—Corporate term for a company (typically a LLC) that produces on-site power and heat from a combined heat and power (CHP) plant, or just electricity from PV systems, for customers under a long-term power purchasing agreement (PPA). From the standpoint of an investor, a Yieldco system "yields" continuous revenue and profit, making it a fungible asset.

Zero Emission Credit (ZEC)—A financial subsidy given to operators of nuclear power plants.

Zero energy building (ZEB)—facility that produces (from renewable sources, typically PV panels) at least as much energy on-site as it consumes. This differs from an "off-the-grid" facility because a ZEB remains connected to the grid and depends on net metering to use its utility to "bank" energy not immediately used by the facility.

Zonal Pricing—Varying wholesale electric rates by Zones.

Zonal Resource Credit (ZRC)—A term first appearing in the MISO (2014), 1 ZRC represents 1 MW of coincident (with the ISO) peak demand capacity for a specific forward year, after discounting to take into account prior forced outages, as defined in the MISO tariff. It is essentially the same concept as Unforced Capacity (UCAP) as used in the NYISO, but in the form of a fungible credit for planning purposes.

Zones—Subsets of a utility's overall territory based on geographical grouping of Nodes.

Index

A

account numbers 139
account rep 36
accounts payable (AP) 47
additional distribution 41
adjustments 101
administrative law judge (ALJ) 103
administrative procedures acts
    106
aggregate 141
aggregated pricing 139
aggregate meters 48
aggregate peak load 65
aggregators 112
air handlers 65, 149
allowable swing 158
Alternative Energy Credits
    (AECs) 91
American Recovery and Rein-
    vestment Act of 2009 62
amperage 21
ancillary services 124
Ariba 136
Association of Energy Engineers
    (AEE) 68, 141
attorney 157
auction commission 136
auditing contract 55
Authorities or Power Marketing
    Administrations (PMA) 6
average electric rate 17

B

backup 53
backup generation 52
balancing 157
bandwidth 126
bidding process 141
bilateral contracts 20, 114
bill, consolidated 121
billing adjustments 54
billing audit 162
bill paying services 48
blackout, 1965 10
blend and extend (B&E) 160
blind bid 137
block 128
block and index 76
BlueLine Innovations 66
brokering commission 125, 136,
    141
brokers 112, 141
Burrell, David 100
business incentives 41, 99
buy-through 99
bypass 86

C

California 113, 117, 155
capacity 21, 26, 124
capacity auction pricing 149
capacity tag 27, 149
carbon footprint 122, 155

carbon neutrality 154
carbon offsets 155
change of electric service 84
CHP deferral rate 54
clamp-on current transformers
    (CTs) 67
coal 121
cogeneration 87
coincident metering 48
coincident peak demand 49
collared 127
co-marketing 153
combined heat and power (CHP)
    9
commission 136, 161
common area loads 61
competitive procurement 111
congestion 124
conjunctive metering 48
connected electric load 59
consolidated 115, 145
consolidated billing 144, 159
consultants 112, 141
consumer advocate 89
consumption 21
contango 145
contract demand 26
contracts for differences (CFD) 129
co-op 58
co-operative utility (co-op) 6, 58
cost of service (COS) 104
cross-subsidized 143
Cunningham, Paul 100
current map of the ISOs 12
curtailment service providers 82
customer 49
customer assets 139
customer charge 29
customer name 160

D
data centers 50, 72, 76, 99, 113
data loggers 67, 77
day-ahead market (DAM) 74,
    146
deferral 37, 78
delivery 12, 20
delivery voltage 35
demand 19
demand control capability 149
demand destruction 78
demand response (DR) 25, 52,
    80, 81, 83, 151
demand response providers 82
dentinstruments.com 67
deposit 37, 144, 159
deregulation 20
descending block rate 49
DHW electric rate 51
diesel generator 53
digital metering 25, 73, 74
digital ("smart") meters 19
discounts 35, 54
discovery process 135
dispute 157
distribution 21, 29
dollar cost averaging 148
due diligence 23, 132
dynamic electric rates 25

E
earth-friendly 122
easement 86
economic development 6, 40, 41,
    46, 99
Ecova 48
EDI 121
efficiency upgrades 56
electronic data interchange (EDI) 47

electronic fund transfer (EFT) 48
emergency generator 53
eminent domain 86, 97
emission generation 76
emission regulations 152
energy 20
energy accounting software 37
energy audit 162
EnergyCAP Express™ 38
EnergyCAP™ 37, 38, 162
EnergyDude™ 38
Energy Information Administration (EIA) 62
energy management system (EMS) 63, 144
Energy Policy Act of 1992 (EP-Act) 10
energy service companies (ESCOs) 53, 96, 153
   contract 78
Energy Watchdog™ 38
Enron 113
environmental attributes 90
Environmental Disclosure Label (EDL) 133
es.eaton.com 31
Excel™ 37
exchanges 116
exemptions 45
exempt wholesale generators (EWG) 10
experimental rate 98
extra utility distribution 56
extra utility equipment 29

F
FAC 22, 29, 31, 41, 56, 126
facilities 59
facilities management 75

Federal Energy Regulatory Commission (FERC) 10
Federal Power Act 86
Feed-in Tariffs (FIT) 91
FERC 86
financial derivatives 57
financial options 57
financial reserve 147
financial risk 112
financial statement 139
Fisher, Roger 100
fixed adder 128
flexible scheduling 76
floating 127
floating price 147
   contract 75
food warehouse 64
force majeure 158
formal complaint 102
fuel adjustment charges (FAC) 22, 29, 31, 41, 56, 126
fuel cells 89, 91
full requirements 52
future pricing 145

G
gas-fired generation 58
generation 21, 29
greenhouse gas (GHG) inventory 162
green tariffs 99, 113
grid operator 65
guaranteed savings 129

H
heat rate 128
hedging 56, 73, 125, 126, 129, 144, 146
high tension 84

high-voltage lines 12
hockey stick hourly pricing 151
hour ahead market or HAM 74
hourly metering 71
hourly power pricing 56, 74, 115
hourly wholesale prices 80
HVAC 45, 119
hydronic heating 63

I
incentive 3, 160
incentive program 56
incumbent supplier 138
independent power producer
    (IPP) 9
independent system operators
    (ISOs) 11
index 127, 159, 162
indicative pricing 133
inflation 48, 160
instantaneous electric load 25
interconnection 13
Intercontinental Exchange 145
interruptible 52
    gas service 151
interval data 62, 68
interval meters 62, 65, 71, 161
intervene 103, 104
intervention 45, 105
investor-owned utility (IOU) 6
Ion Wave Technologies 136
ISO or transmission charges 125,
    149

K
kilowatt capacity (kW) 19
kilowatt-hour (kWh) 21
kilowatt (kW) 21
kVAR (kilovolt-amps reactive) 77

kW-year 81

L
laboratory 65
laddering 147
landlord 60, 61
lawsuit 106
letter of credit (LoC) 145, 147
licensed retailers/suppliers 117
lighting 119
line loss 41, 125, 152
liquidated damages 159
load factors (LF) 71, 143
load management 52, 76
    options 83
load profile 62, 63, 70, 73
load shape 62
load survey 59
local distribution companies
    (LDC) 13
low emission 91
low tension 84

M
majeure 157
management 59
mandatory TOU rates 70
manufacturing association 46
material change 158
material deviation 158
medical research 99
Megawatt Daily 144
metering 48
metering/billing 29
Metrix™ 37, 38, 162
Michigan 117
microturbine 53
mill 22
mission statement 122

municipalization 97
municipal utility (muni) 6, 58
MW-day 81

N
national RECs 155
natural gas 119
natural gas-driven systems 78
Nebraska 20
negotiation 97, 99
net metering 89, 90
New England 113
New York 113
nitrous oxide (NO$_x$) 53
non-electric chillers 72
non-profits 45, 103
non-utility generators (NUG) 9
non-utility suppliers 20
North American Electric Reliability Council (NERC) 10
North American Energy Standards Board (NAESB) 157
North American Industrial Classification System (NAICS) 51
NV 113
NY 111
NY ISO 80

O
occupancy sensors 23
off-site renewable power 113
off-tariff discounts 118
off-tariff rates 42
on-bill payment 85
online auction software 136
online reverse auctioneer 141
online reverse auctions 135
on-peak 21
onsetcomp.com 67

on-site CHP 72
on-site generation 90
on-site non-utility supplier 87
operating lease 154
Oregon 131
organized markets 12, 74, 131
outstanding late fees 160

P
packages 150
pass-through 57
peak demand 21, 24, 26, 64
peak kW demand 3
penalties 37, 127
PF correction capacitors 77
photovoltaic (PV) power 15
pilot 98
  rates 42
PJM Interconnection 82
Polar Vortex winter 74
power 21
power analyzer 77
power factor (PF) 21, 22, 29, 30, 77
power pool/ISO 32
power purchasing agreement (PPA) 87, 114
power storage 90, 152
Power To Choose 117
price discovery 138
price risk 129, 148
price to beat 115
price to compare 115
price volatility 146, 147
pricing spreadsheet 139
pricing structures 112
primary 84
primary demand charge 85
proceedings 103, 104, 106

program charges 29, 41
property 49
provider of last resort 115
public hearings 105
publicly-owned 5
Public Utility Commissions
    (PUCs) 6
Public Utility Regulatory Policies
    Act (PURPA) 9
pulse meters 24, 67
purchase of receivables 146
purchasing 59, 75

Q

qualifying facility (QF) 9
quid pro quo 86

R

rack rates 71
ratchet 22, 26
    charge 27
rate class 21, 35
rate-of-return 23
reactive power 30
    charge 77
real time/hour-ahead market (RT
    or HAM) 146
real-time pricing (RTP) 73
recession 76
regional transmission operators
    (RTO) 12
religious 45
religious property 45
renewable energy 136
Renewable Energy Certificates
    (RECs) 90
renewable energy credits (RECs)
    154
renewable portfolio standard

(RPS) 125, 90
renewable power 15
renewal 157
request for information (RFI) 118
request for proposals (RFP) 114,
    134
resiliency 154
retail deregulation 13, 112
retail electricity providers 117
retail power suppliers 112
return-on-equity (ROE) 8
revenue neutral 72
revenue requirement 104
reverse auctions 138
riders 99
right of first refusal 160
RPS 155

S

sales tax 32, 46, 111
sample contracts 136
sealed bid 136
secondary demand charge 85
secondary distribution 84
service addresses 139
settlement 105
site 49
smart meters 62, 115
solar photovoltaic (PV) 89, 119
Southern Power Pool (SPP) 11
special conditions 35
specialists 112, 114, 116
special provisions 99
specifications 77
spreadsheet programs 36
Standard Industrial Classifica-
    tions (SIC) 51
standard offer 115, 116
standard offers 112

strap-on meter 66
Studebaker, John 88, 100
sub-meters 46, 60, 61
subsidies 32
substation 84, 86
summary billing 47, 121
supplier's adder 127
supply 12, 19
supply risk 126
surcharges 32
sustainability 120, 121
swing 126, 144
switchover 114
system benefit charge 32

T
tariffs 3, 8, 17, 18, 20, 23, 35, 36, 42, 60
tax-deductible donation 161
taxes 29, 126
tax-exempt 139
tax status 35
tenant 59
term 144
termination penalty 159
Texas 74, 146
thermal meter 24
thermal storage 70, 72
tiered blocks 30
timed "blind" bids 135, 138
time-of-use (TOU) 19, 61, 66, 68
time-sensitive tariffs 25
tolling 58
    agreement 58
tranches 162
transformer 8, 9, 85
transition charge 32

transmission 21, 29
transmission bypass 54, 86

U
unbundling 20, 116
upgrade financing 153
Ury, William 100
utility bill auditor 54
utility distribution companies (UDC) 13
utility tax 32

V
value engineering 78
variable speed drives (VSD) 65, 78, 149
Vermont 113
volatile pricing 73
voltage 21, 83, 84
    reduction 85
    upgrade 162
volt-amps-reactive, or VAR 30
volts 13

W
WattVision 66
wheeling 15, 99, 113, 154
wholesale market 14
    pricing 127
wind 111, 121, 154
wireless and web-based metering systems 60
word processing 36
writ of mandamus 106
www.dsireusa.org 89
Wyoming 155

Printed in the United States
by Baker & Taylor Publisher Services